Springer Complexity

Springer Complexity is an interdisciplinary program publishing the best research and academic-level teaching on both fundamental and applied aspects of complex systems—cutting across all traditional disciplines of the natural and life sciences, engineering, economics, medicine, neuroscience, social and computer science.

Complex Systems are systems that comprise many interacting parts with the ability to generate a new quality of macroscopic collective behavior the manifestations of which are the spontaneous formation of distinctive temporal, spatial or functional structures. Models of such systems can be successfully mapped onto quite diverse "real-life" situations like the climate, the coherent emission of light from lasers, chemical reaction-diffusion systems, biological cellular networks, the dynamics of stock markets and of the Internet, earthquake statistics and prediction, freeway traffic, the human brain, or the formation of opinions in social systems, to name just some of the popular applications.

Although their scope and methodologies overlap somewhat, one can distinguish the following main concepts and tools: self-organization, nonlinear dynamics, synergetics, turbulence, dynamical systems, catastrophes, instabilities, stochastic processes, chaos, graphs and networks, cellular automata, adaptive systems, genetic algorithms and computational intelligence.

The three major book publication platforms of the Springer Complexity program are the monograph series "Understanding Complex Systems" focusing on the various applications of complexity, the "Springer Series in Synergetics", which is devoted to the quantitative theoretical and methodological foundations, and the "Springer Briefs in Complexity" which are concise and topical working reports, case studies, surveys, essays and lecture notes of relevance to the field. In addition to the books in these two core series, the program also incorporates individual titles ranging from textbooks to major reference works.

Series Editors

Understanding Complex Systems

Founding Editor: S. Kelso

Future scientific and technological developments in many fields will necessarily depend upon coming to grips with complex systems. Such systems are complex in both their composition—typically many different kinds of components interacting simultaneously and nonlinearly with each other and their environments on multiple levels—and in the rich diversity of behavior of which they are capable.

The Springer Series in Understanding Complex Systems series (UCS) promotes new strategies and paradigms for understanding and realizing applications of complex systems research in a wide variety of fields and endeavors. UCS is explicitly transdisciplinary. It has three main goals: First, to elaborate the concepts, methods and tools of complex systems at all levels of description and in all scientific fields, especially newly emerging areas within the life, social, behavioral, economic, neuro- and cognitive sciences (and derivatives thereof); second, to encourage novel applications of these ideas in various fields of engineering and computation such as robotics, nano-technology and informatics; third, to provide a single forum within which commonalities and differences in the workings of complex systems may be discerned, hence leading to deeper insight and understanding.

UCS will publish monographs, lecture notes and selected edited contributions aimed at communicating new findings to a large multidisciplinary audience.

More information about this series at http://www.springer.com/series/5394

Nikolai Verichev · Stanislav Verichev ·
Vladimir Erofeev

Chaos, Synchronization and Structures in Dynamics of Systems with Cylindrical Phase Space

 Springer

Nikolai Verichev
Mechanical Engineering Research Institute
of the Russian Academy of Sciences
Nizhny Novgorod, Russia

Stanislav Verichev
Allseas Engineering B.V.
Delft, Zuid-Holland, The Netherlands

Vladimir Erofeev
Mechanical Engineering Research Institute
of the Russian Academy of Sciences
Nizhny Novgorod, Russia

ISSN 1860-0832   ISSN 1860-0840 (electronic)
Understanding Complex Systems
ISBN 978-3-030-36105-1   ISBN 978-3-030-36103-7 (eBook)
https://doi.org/10.1007/978-3-030-36103-7

This Springer imprint is published by the registered company Springer Nature Switzerland AG
The registered company address is: Gewerbestrasse 11, 6330 Cham, Switzerland

Preface

Systems of physical pendulums [1, 2], systems of superconductive junctions [3–5], systems of coupled electric machines [6] and vibrational mechanisms [7–9], phase synchronization systems [10, 11]—this list is far from a complete set of systems from various areas of natural science and technical applications that have a common property: some physical variables that determine the dynamic state of these systems are space-periodic, or, in other words, angular or phase variables. The dynamics of such systems are described by similar mathematical models defined in cylindrical phase space, i.e., by the dynamical systems of coupled rotators of the form

$$I\ddot{\varphi}_k + \delta_k(1 + F_{1k}(\varphi_k))\dot{\varphi}_k + \sigma_k F_{2k}(\varphi_k) = \gamma_k + F_k(\boldsymbol{\varphi}, \dot{\boldsymbol{\varphi}}, \boldsymbol{\psi}, z),$$

$$\dot{\mathbf{z}} = \mathbf{A}\mathbf{z} + \mathbf{Z}(\mathbf{z}, \boldsymbol{\varphi}, \dot{\boldsymbol{\varphi}}, \boldsymbol{\psi}), \tag{1}$$

$$\dot{\psi} = \omega_0.$$

Here, $k = \overline{1, n}$; $\mathbf{A}$ is $(m \times m)$ constant stable matrix; I, $\delta_k, \sigma_k, \gamma_k$ and ω_0 are the constant dimensionless positive parameters. Variables φ, ψ are the phase (angular) variables, such that $\varphi_k = \varphi_k + 2\pi(\mathrm{mod}2\pi)$, $\psi = \psi + 2\pi(\mathrm{mod}2\pi)$, $z \in R^m$. The differentiation is done over a dimensionless time τ. Functions $F_{1k}(\varphi_k)$, $F_{2k}(\varphi_k)$ have zero mean values $\langle F_{1k}(\varphi_k)\rangle_{\varphi_k} = 0$, $\langle F_{2k}(\varphi_k)\rangle_{\varphi_k} = 0$, and functions of couplings $F_k(\boldsymbol{\varphi}, \dot{\boldsymbol{\varphi}}, \boldsymbol{\psi}, z)$, $\mathbf{Z}(\boldsymbol{\varphi}, \dot{\boldsymbol{\varphi}}, \boldsymbol{\psi})$ are periodic with respect to the phase variables. System (1) is defined in cylindrical phase space $G(\boldsymbol{\varphi}, \dot{\boldsymbol{\varphi}}, \boldsymbol{\psi}, z) = T^{n+1} \times R^{n+m}$.

The unity of mathematical models allows one to consider not individual physical systems, but specific models of the form (1) as independent mathematical objects with a subsequent interpretation of the results onto different physical systems. On the other hand, translation of the dynamics of models into their phase spaces allows not only adequately describing their nonlinear properties, but also using the entire set of methods of the qualitative theory of differential equations, the theory of bifurcations, and qualitative-numerical methods of investigation. In this context, studies of the dynamics of systems of the form (1) are carried out, and the interpretation of results is provided. This idea, which eventually has grown into a method of modern science, belongs to A. A. Andronov and was outlined in the pioneering book "Theory of Oscillations" (by A. A. Andronov, A. A. Vitt, and S. E. Khaikin) that was published in the middle of the last century. Many years later, according to the rule of chance, this idea has returned to the scientific world as a "new discovery" and has been called the synergetics.

This monograph consists of six chapters and is built according to the principle "from simple to complex." Chapter 1 provides an overview of the properties of a model with one degree of freedom and exemplifies physical systems governed by this model. The basic definitions are introduced, the subject of research is concretized, and the approbation of methods is given.

Chapter 2 is devoted to the dynamics of systems with one and a half degrees of freedom. These models describe certain devices of superconductor electronics, electric motors loaded by an unbalanced aperiodic load, and other systems. The study in this chapter emphasizes on strange attractors, leading to chaotic rotation of the rotators.

Chapter 3 explores the dynamic properties of models with two degrees of freedom. Qualitative pictures of rotation characteristics of rotators, as well as resonance characteristics of vibrational systems acting as loads, are presented. It is pointed out that there exist regions of the instability of these characteristics caused by strange attractors such as Lorentz and Feigenbaum attractors. Interpretation of results on actual physical systems is given.

Chapter 4 deals with the vibrations of rotating shafts represented in the framework of models of the form (1). It is shown that, in addition to the classical Sommerfeld effect, the dynamics of the shafts can face the chaotization of torsional vibrations. The method of damping of the bending vibrations of rotating shafts by means of controlling the speed of the actuator is described in this chapter.

In Chap. 5, we discuss the synchronization in lattices of dynamical systems modeled by a chain, two-dimensional and three-dimensional lattices, as well as by a ring of identical coupled systems. The conditions for local and global stability of isochronous synchronization are presented.

Chapter 6 is devoted to the cluster dynamics of a chain and of a ring of dynamical systems. The necessary terms and definitions are introduced, and a set of equivalent transformations and rules for the synthesis of schemes of cluster structures are provided. Classification of cluster structures is given. A number of theorems regarding a number of types of cluster structures in a chain and in a ring are proven. The results of the investigation of the stability of cluster structures in lattices of this type are presented.

<table>
<tr><td>Nizhny Novgorod, Russia</td><td>Nikolai Verichev</td></tr>
<tr><td>Delft, The Netherlands</td><td>Stanislav Verichev</td></tr>
<tr><td>Nizhny Novgorod, Russia</td><td>Vladimir Erofeev</td></tr>
</table>

References

1. Andronov, A.A., Vitt, A.A., Khaikin, S.E.: Theory of Oscillators. Dover Publications, New York (1987)
2. Barbashin, E.A., Tabueva, V.A.: Dynamic Systems with a Cylindrical Phase Space. Nauka, Moscow (1969) (in Russian)
3. Josephson, B.D.: The discovery of tunnel supercurrents. Rev. Mod. Phys. **46**, 252–254 (1974)
4. Barone, A., Paterno, G.: Physics and Applications of the Josephson Effect. Wiley, NewYork (1982)
5. Likharev, K.K.: Dynamics of Josephson Junctions and Circuits. Gordon and Breach, New York (1986)
6. Gorev, A.A.: Selected Works on Questions of the Stability of Electric Systems. Gosenergoizdat, Moscow (1960) (in Russian)
7. Blekhman, I.I.: Synchronization of Dynamical Systems. Nauka, Moscow (1971) (in Russian)
8. Alifov, A.A., Frolov, K.V.: Interaction of Non-Linear Oscillatory Systems with Energy Sources. Taylor & Francis (1990)
9. Adams, M.L.: Rotating Machinery Vibration: From Analysis To Troubleshooting. Marcel Dekker (2000)
10. Akimov, N., Belyustina, L.N., Belykh, V.N., et al.: Phase Synchronization Systems. Radio Svyaz', Moscow (1982) (in Russian)
11. Lindsey, W.C.: Synchronization Systems in Communication Control. Prentice-Hall, Inc., Englewood Cliffs, New Jersey (1972)

Contents

About the Authors

Nikolai Verichev is Principal Researcher at Mechanical Engineering Research Institute of the Russian Academy of Sciences. He graduated from Nizhny Novgorod State University in 1974 (M.Sc. in Radio Engineering). He got his Ph.D. degree in Nizhny Novgorod State University (1985). He specializes in the field of the qualitative theory of dynamical systems, oscillation theory, nonlinear dynamics, dynamical chaos, and chaotic synchronization. He has discovered the effect of chaotic synchronization of non-identical systems (1985). He is author/co-author of 29 papers, two books, and two patents. e-mail: nverichev@yandex.ru

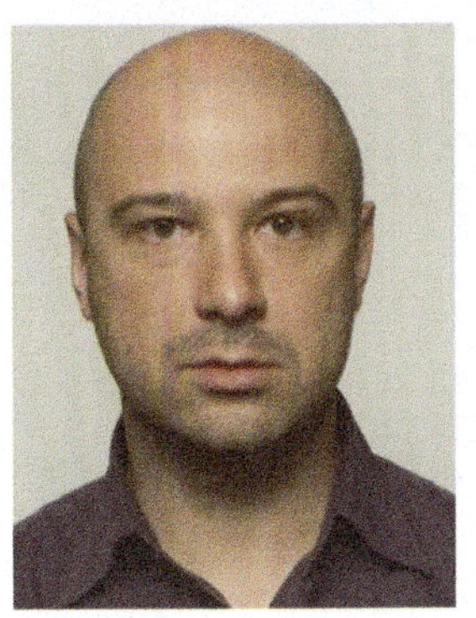

Stanislav Verichev is Senior R&D Engineer at Allseas. He graduated from Nizhny Novgorod State University in 1998 (M.Sc. in Physics). He got his Ph.D. degree in TU Delft (2002). He specializes in nonlinear dynamics as well as such practical fields as dredging and mining, oil and gas, civil and mechanical engineering, and deep sea mining. He is author/co-author of 36 papers, four books, and eight patents. e-mail: s.verichev@mail.ru

Vladimir Erofeev is Professor, Director of Mechanical Engineering Research Institute of the Russian Academy of Sciences. He got his Ph.D./Dr. Habil. degrees in Nizhny Novgorod State University (1986, 1994). He has proposed new methods of non-destructive testing of materials and structures. He developed scientific foundations of vibration protection systems for machines and structures using inertia and dissipation of rheological media. He performed a series of studies on the chaotic dynamics of mechanical systems with energy sources of limited power, as well as on the existence and stability of stationary cluster structures in homogeneous chains of dissipative coupled rotators. e-mail: erof.vi@yandex.ru

Chapter 1
Autonomous and Non-autonomous Systems with One Degree of Freedom

1.1 Dynamics of Autonomous Rotator and Related Physical Systems

A simplest model of the form (1) can be exemplified by autonomous rotator, which is governed by equation of the form

$$I\ddot{\varphi} + (1 + F_1(\varphi))\dot{\varphi} + F_2(\varphi) = \gamma. \tag{1.1}$$

Dynamical properties of this equation for the various nonlinear functions have been studied in numerous papers and are currently well known [1–3]. Let us consider the following case: $F_1(\varphi) = \varepsilon \cos \varphi$, $F_2(\varphi) = \sin \varphi$, $|\varepsilon| \leq 1$.

Figure 1.1 shows bifurcation diagram (Fig. 1.1a) and phase portraits for Eq. (1.1) on the involute of the phase cylinder $(\varphi, \dot{\varphi})$ for each of the parameter domains, see Fig. 1.1b–d (all points of the phase portrait on the left and right boundaries of the involute are identical). On the bifurcation diagram: $\lambda = I^{-1/2}$, $\lambda^*(\gamma)$ is a bifurcation curve (Tricomi curve) of the separatrix saddle loop (saddle knot for $\lambda > \lambda_0$), at the intersection of which, from left to right, a limit cycle is generated from the loop, and when it is done in the opposite direction, the limit cycle "sticks" into the loop. Bifurcation line $\gamma = 1$ corresponds to merging of equilibria with their further disappearance if the intersection of that line occurs from left to right; or to birthing of equilibria if parameter γ is varied in the opposite direction.

Phase portraits (Fig. 1.1b–d): O_1 is a stable equilibrium of focus or knot type, O_2 is an equilibrium of the saddle type, and L is a stable limit cycle, which is born from the separatrix loop of the saddle (or of the saddle knot) during the reverse variation of parameter γ (from left to right) and which "sticks" into this loop for the reverse variation of parameter γ. The limit cycle moves upwards along the cylinder (the speed of rotation increases) with increasing parameter γ. The equilibria converge as the approximation of this parameter approaches the line $\gamma = 1$ from the left.

© Springer Nature Switzerland AG 2020

N. Verichev et al., *Chaos, Synchronization and Structures in Dynamics of Systems with Cylindrical Phase Space*, Understanding Complex Systems,
https://doi.org/10.1007/978-3-030-36103-7_1

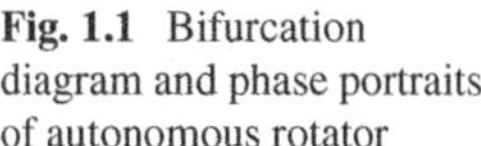
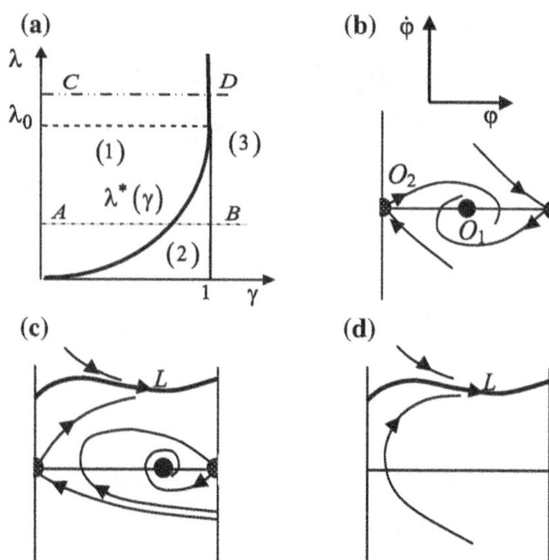

Fig. 1.1 Bifurcation diagram and phase portraits of autonomous rotator

The parameter domain (1) shown in Fig. 1.1a corresponds to the global stability of equilibrium O_1 shown in Fig. 1.1b. In this case, the rotator (imagine a pendulum) in the course of time comes from any initial state to a given state of equilibrium (for a pendulum it would be the lower one). For parameters from parameter domain (2) (see Fig. 1.1a), in addition to the stable equilibrium, there exists a rotational limit cycle (see Fig. 1.1c). In this case, under certain initial conditions, the rotator comes to a state of equilibrium, while for other ones it goes to the stationary rotational regime. For parameters from parameter domain (3) (see Fig. 1.1a), the limit cycle is stable globally (see Fig. 1.1d). For these parameters, the rotator in the course of time goes from any initial state to the stationary rotational mode.

As already said, for transition from the parameter domain (1) into domain (2), a limit cycle is born from the separatrix loop of saddle O_2 while for transition from domain (1) into domain (3), it is born from the separatrix loop of a saddle knot. Bifurcation values of parameters that correspond to the separatrix loop belong to the Tricomi curve $\lambda = \lambda^*(\gamma)$. The rotational frequency of the limit cycle increases from zero as the torque γ increases. In this case, on the phase cylinder, the limit cycle moves upwards. Line $\gamma = 1$ corresponds to a complex equilibrium, which is formed from merging of saddle O_2 and focus (knot) O_1. This equilibrium disappears when parameters cross this line to move into parameter domain (3). Phase portraits for bifurcation values of parameter γ are not shown here. To complete the picture, one could add the second part of the bifurcation diagram (see Fig. 1.1a), which is symmetrical with respect to the ordinate axis. For that part of the diagram, the same bifurcations will take place, just occurring on the lower part of the phase cylinder. This follows from the invariance of Eq. (1.1) to the substitution $\gamma \to -\gamma,\ \varphi \to -\varphi$.

Let us introduce a rotation characteristic of rotator as a function providing the "integral" information about its dynamics when the rotator interacts with loads as well as other rotators in a coupled system.

Definition The function $\Omega = \lim_{T \to \infty} \frac{1}{T} \int_0^T \dot{\varphi}(\tau, \tau_0) d\tau$, defined under the parameter space of system (I1) and the space of its initial conditions, is called the rotation characteristic of rotator.

As one can see, rotation characteristic represents dependence of the average rotational frequency of rotator on the system parameters and its initial conditions.

Any solution of system (I1) for the time derivative of the phase, which represents an instant rotational frequency, can be written in the following form: $\dot{\varphi}(\tau, \tau_0) = \widetilde{\dot{\varphi}}(\tau, \tau_0) + \dot{\varphi}^*(\tau, \tau_0)$, where $\widetilde{\dot{\varphi}}$ is a solution that corresponds to the transient process towards the final solution $\dot{\varphi}^*$.

I.e. $\dot{\varphi}^*(\tau, \tau_0) = \lim_{\tau \to \infty} \dot{\varphi}(\tau, \tau_0)$, $\lim_{\tau \to \infty} \widetilde{\dot{\varphi}}(\tau, \tau_0) = 0$ (the value of $\left|\widetilde{\dot{\varphi}}(\tau, \tau_0)\right|$ has a meaning of deviations of solutions $\dot{\varphi}(\tau, \tau_0)$ and $\dot{\varphi}^*(\tau, \tau_0)$).

Using the latter expressions, we obtain:

$$\Omega = \lim_{T \to \infty} \frac{1}{T} \int_0^T \dot{\varphi}^*(\tau, \tau_0) d\tau.$$

Solutions $\dot{\varphi}^*(\tau, \tau_0)$ can correspond to equilibria, limit cycles, invariant tori, and strange attractors of system (I1).

It can be seen from the definition that the problem of obtaining different qualitative forms of rotation characteristics for different system parameters is related to a classical problem of the decomposition of the parameters' space into the domains corresponding to the qualitatively different structures of trajectories in the phase space of the system.

For an autonomous rotator, solutions $\dot{\varphi}^*(\tau, \tau_0)$ of system (1.1) correspond to equilibria and limit cycles. Let us build qualitative forms of rotation characteristic of autonomous rotator for variable parameter γ using the aforementioned information regarding the dynamics of the model (according to its phase portraits, see Fig. 1.1). Traditionally, for the sake of convenience, the inversed function $\gamma = \gamma(\Omega)$. will be plotted.

Suppose that parameter λ corresponds to point C, and parameter γ itself varies along CD from zero. During such parameter variation, first there exists a globally asymptotically stable equilibrium (see Fig. 1.1b), which is the case for $0 \le \gamma \le 1$. At the same time $\Omega = 0$, which defines a so-called zero branch of the rotation characteristic. For $\gamma = 1 + 0$, equilibrium disappears and a limit cycle is born from a separatrix loop of saddle-focus, which has extremely small frequency $\Omega = 0 + 0$ (see Fig. 1.1d). The pendulum starts to rotate. The frequency of rotation increases from zero as the moment γ increases, which corresponds to the upward motion of the limit cycle along the phase cylinder. Rotational branch L of rotation characteristic corresponds to the limit cycle. Bisectrix $\gamma = \Omega$ is an asymptote for this curve, as can be seen directly from Eq. (1.1): $\dot{\varphi} \to \gamma$, if $\gamma \to \infty$. Because of the invariance

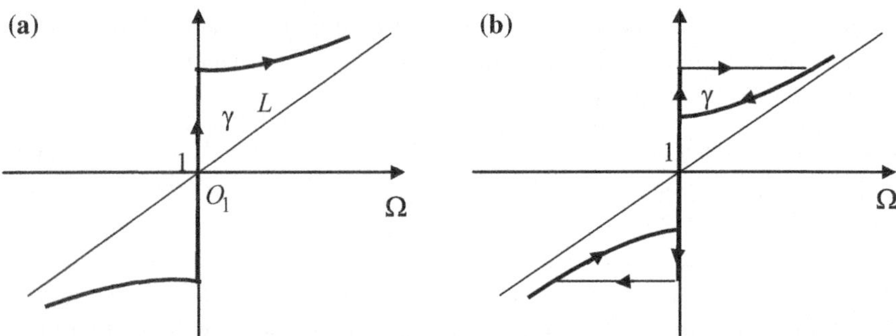

Fig. 1.2 Rotation characteristic for varied parameter γ along line CD (**a**) and along line AB (**b**)

of the equation to change $\varphi \to -\varphi$, $\gamma \to -\gamma$, the same is observed for the lower half-plane (Ω, γ). In general, a qualitative form of rotational characteristic is shown in Fig. 1.2a.

Suppose now that parameter λ corresponds to point A, and parameter γ is varied along line AB. As in the previous case, the rotation characteristic has a zero branch defined by an equilibrium that exists in the interval $0 \leq \gamma \leq 1$. A qualitative difference, however, consists of the following: while γ is increased within this interval, a stable rotational limit cycle is born (see Tricomi curve) and along with the equilibrium there is also a possibility of rotations. At the same time, by the moment of time when equilibrium disappears, the limit cycle moves upwards along the phase cylinder and acquires a determined frequency of rotation $\Omega > 0$. This results in the following: for $\gamma = 1 + 0$, there is a jump in the frequency of rotation from the zero branch to the rotational branch of rotation characteristic. For a further increase of γ, rotational branch of rotational characteristic keeps behaving in the same way. On the contrary, for a decrease of parameter γ, the rotator "stays" on a limit cycle and, until it disappears, does not "realize" that equilibrium is born. The frequency of rotation of the limit cycle decreases and becomes zero for $\gamma = \lambda^*$, when the limit cycle gets "stuck" in the separatrix loop of the saddle. In this case, the rotational branch adjoins the zero branch of the rotation characteristic. An explicit hysteretic loop takes place on the rotational characteristic. The complete picture of the hysteretic rotational characteristic, taking into account a symmetric part of the bifurcation diagram, is shown in Fig. 1.2b. From the bifurcation diagram, one can see that these two types represent all possible types of qualitative forms of the rotational characteristic of an autonomous rotator.

For a detailed study of the rotational motions of rotators, the methods of small parameter will be used [4–7], as well as the method of point mapping [8] closely related to the method of the averaging. The latter will be applied on the basis of special transformations of systems from the class (I1) that are outlined in Appendix I.

Let us consider in more detail the rotational motions of the rotator, starting with the demonstration of the algorithm for transforming the rotator's equation into the so-called standard form.

The problem: to transform Eq. (1.1) to an equivalent system of the form

$$\dot{\xi} = \mu \, \Xi(\xi, \varphi),$$
$$\dot{\varphi} = \omega + \mu \Phi(\xi, \varphi), \tag{1.2}$$

where ξ is a certain new variable, $\mu = I^{-1}$, ω is a parameter that will be defined later, and $\Xi(\xi, \varphi)$ and $\Phi(\xi, \varphi)$ are bounded by φ (periodic) functions that will also be defined later.

The solution: we substitute the second equation of system (1.2) into Eq. (1.1) to obtain:

$$\frac{\partial \Phi}{\partial \varphi}(\omega + \mu \Phi) + \frac{\partial \Phi}{\partial \xi}\dot{\xi} + (1 + F_1(\varphi))(\omega + \mu \Phi) + F_2(\varphi) = \gamma. \tag{1.3}$$

We exclude parameter ω and unknown functions from this equation, taking into account the natural condition of the boundedness of the sought functions with respect to the phase variable φ and required form of system (1.2). From the condition of the boundedness, we obtain that $\omega = \gamma$. Taking into account the form of the second equation of system (1.2), we obtain an equation defining function $\Phi(\xi, \varphi)$ (by separating terms that do not have multiplier μ):

$$\frac{\partial \Phi}{\partial \varphi}\omega + \omega F_1(\varphi) + F_2(\varphi) = 0. \tag{1.4}$$

The simplest solution for Eq. (1.4) has the form:

$$\Phi(\varphi, \xi) = -\int F_1(\varphi)d\varphi - \frac{1}{\omega}\int F_2(\varphi)d\varphi + \xi.$$

The remaining terms in (1.3) determine the equation for the function $\Xi(\xi, \varphi)$:

$$\Xi(\xi, \varphi) = -\Phi\frac{\partial \Phi}{\partial \varphi} - (1 + F_1(\varphi))\Phi.$$

In the case when $F_1(\varphi) = \varepsilon \cos \varphi$, $F_2(\varphi) = \sin \varphi$, one obtains:

$$\dot{\xi} = \mu\left(-\varepsilon \sin \varphi + \frac{1}{\omega}\cos \varphi + \xi\right)\left(1 - \frac{1}{\omega}\sin \varphi\right),$$
$$\dot{\varphi} = \omega + \mu\left(-\varepsilon \sin \varphi + \frac{1}{\omega}\cos \varphi + \xi\right). \tag{1.5}$$

Note that system (1.5) is equivalent to Eq. (1.1). On the other hand, if μ is a small parameter (usually, the level of smallness of the parameter in asymptotic methods is not discussed, although there exist papers devoted to its evaluation), then system (1.5) has a "standard form of a system with a rapidly rotating phase" (φ). The technique of averaging such systems has been developed in full detail in [9] and, for that reason, is not discussed here. Averaging (1.5) to terms of order μ^3, we obtain an averaged system of the form

$$\dot{\xi} = -\mu\left(\frac{\varepsilon}{2\omega} + \xi\right) + \mu^2\left(\frac{\varepsilon^2}{2\omega} + \frac{\varepsilon}{2\omega^2}\xi\right),$$

$$\dot{\varphi} = \omega + \mu\xi - \mu^2\left(\frac{\varepsilon^2}{2\omega} + \frac{1}{2\omega^3}\right). \tag{1.6}$$

Equations (1.6) retain the previous notations for the averaged variables.

It is known that a limit cycle of system (1.2) corresponds to the equilibrium of system (1.6) and, respectively, the one of Eq. (1.1) including the conditions of stability. As one can see, the equilibrium is unique and stable. Its coordinate $\xi_0 = \frac{-\frac{\varepsilon}{2\omega}(1-\mu\varepsilon)}{1-\frac{\mu\varepsilon}{2\omega^2}}$. After substituting this expression into the second equation of system in (1.6), we obtain an expression for the rotational branch of rotation characteristic in an explicit form (of the order of μ^3):

$$\Omega = \gamma - \frac{\mu\varepsilon}{2\gamma} - \frac{\mu^2}{4\gamma^3}\left(\varepsilon^2 + 2\right). \tag{1.7}$$

Figure 1.3 shows the analytical result of (1.7) compared to results of numerical modeling [10]. One can see that even for not small values of parameter μ $\left(\mu = 1/2\right)$, there is good qualitative and quantitative matching (a so-called "miracle" of a small parameter).

Comment. Further, we will be interested in the qualitative features of the dynamics of systems, so we will not carry out the averaging for orders higher than the first one. The right-hand sides of systems averaged in the first approximation represent mean values of the corresponding right-hand sides taken over the period of a fast

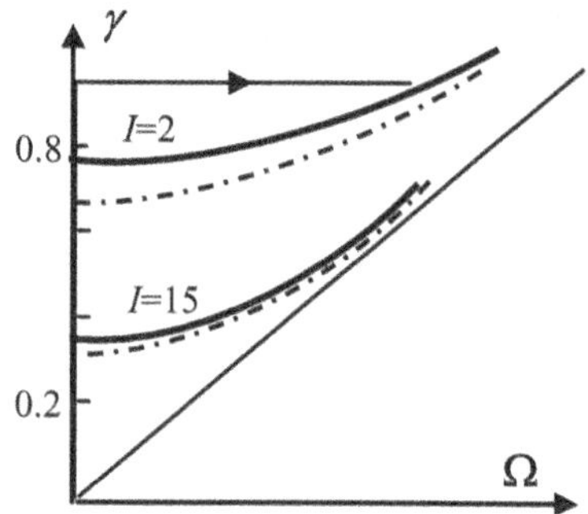

Fig. 1.3 Rotation characteristic: bold lines indicate results of numerical modeling, while the dashed lines show analytical results

Fig. 1.4 Froude pendulum

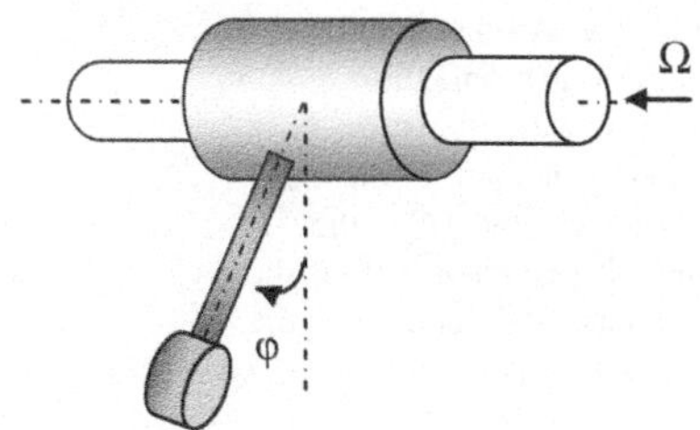

phase. Such truncated systems contain all of the interesting qualitative features of the original systems. The justification for this is the term "coarse dynamical system" (structurally stable dynamical system) introduced by Andronov [1, 11].

Now let us consider examples of physical systems, the dynamics of which is modeled by the rotator equation.

Example 1 Froude pendulum [12], see Fig. 1.4. The Froude pendulum represents a shaft rotating with a constant angular speed Ω inside a sleeve. There is viscous damping in between the shaft and the sleeve. The moment from the friction force acting on pendulum has the form $\delta(\Omega - \dot{\varphi})$, the moment from the gravity force has the form $mgl \sin \varphi$, and ml^2 is the moment of inertia of the pendulum. The equation of motion has the form:

$$ml^2\ddot{\varphi} + \delta\dot{\varphi} + mgl \sin \varphi = \delta\Omega.$$

After the transformation of time and renaming of parameters, this equation can be transformed to Eq. (1.1). In this case, the rotation characteristic has a literal meaning of the dependence of the angular speed of the pendulum's rotation on the external driving moment, which is proportional to the speed of rotation of the shaft. In principle, all features of the dynamics of this system have already been discussed above.

Example 2 Synchronous electric machine. Generally, this physical system is governed by a six-dimensional model [13, 14]. However, due to the fact that magnitudes of changes of electric and mechanical variables have different orders, the motion of the rotor can be governed by a dynamical system of the form [3]:

$$I\ddot{\varphi} + (\alpha + \beta \cos 2\varphi - \delta \sin \varphi)\dot{\varphi} + \sin \varphi = T.$$

Here I is the moment of inertia of the rotor; α, β, and δ are the coefficient of viscous damping and two coefficients of electric damping, respectively; T is the constant moment acting out of the load; and φ is the angle between magnetic fields of the rotor and stator. As one can see, accurate within designations, this equation represents the equation of rotator (1.1).

Example 3 System of phase synchronization [15, 16], see Fig. 1.5.

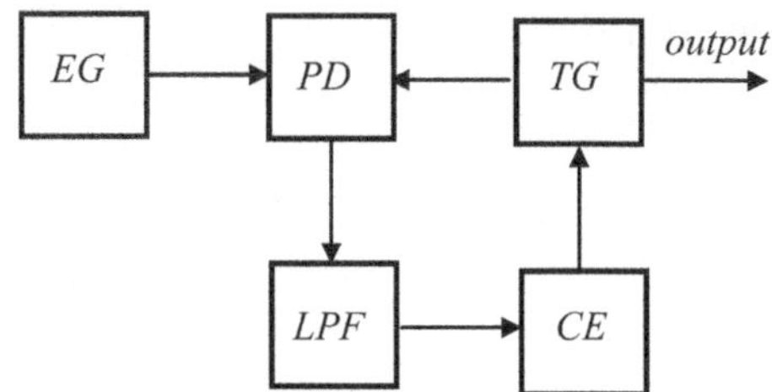

Fig. 1.5 A scheme of a phase synchronization system. Here EG is the etalon generator, PD is the phase detector, TG is the tunable generator, LPF is the low-pass filter, and CE is the control element

In the case of a sinusoidal characteristic of the phase detector (PD) and for a transfer ratio of the low-pass filter (LPF) $K(p)$, (with p the operator of differentiation), this system is governed by equation of the form

$$p\varphi + \Omega_{hr} K(p) \sin \varphi = \Omega_{im}.$$

Here Ω_{hr}, Ω_{im} are the holding range and initial mistuning of etalon and tunable generators, and φ is their phase difference. In the case of the first-order filter of the form $K(p) = (T_1 p + 1)/(T p + 1)$, where T_1, T are constants of the filter, simple transformations reduce this equation to the form (1.1).

Example 4 Superconductive Josephson junction [10, 17, 18]. In 1962, B. Josephson theoretically predicted the existence of two phenomena within dynamical behavior of the junction of two superconductors [17] (see Fig. 1.6). The first phenomenon represents a possibility of existence of surge current I_s that does not cause a voltage drop at the junction.

The second phenomenon consists in the following: if a there exists a voltage V at the junction, then its value is related to the frequency of the electromagnetic field via fundamental constants: $\omega = 2eV/h$ (Josephson relation), where e and h are the electron charge and Planck's constant. In the simplest case $I_s = I_c \sin \varphi$, where φ is the phase difference of phases of the quantum-mechanical wave functions of the parameter of order of superconductors [17]. The dynamical (resistive) model is

(a)   **(b)**   **(c)**

Superconductors

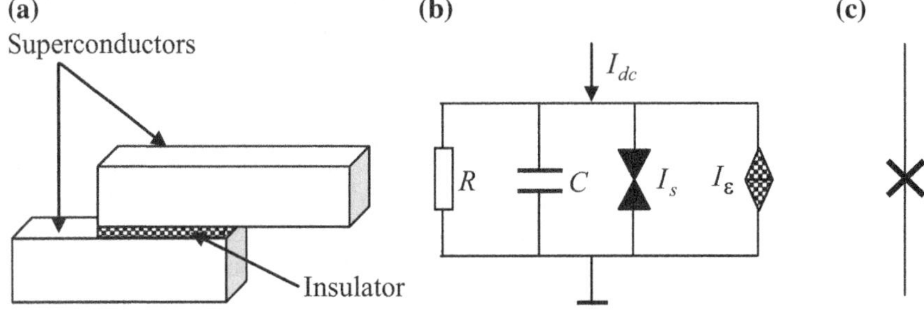

Fig. 1.6 Superconductive junction (**a**), electric scheme of resistive model (**b**), schematic representation (**c**)

introduced in [19]. According to this model, an equivalent scheme of the junction with determined external current I represents a parallel coupling of the capacitor, resistor, and surge generator (see Fig. 1.6b).

Josephson indicated one more component of the electric current through the junction: $I_\varepsilon = \varepsilon V \cos \varphi$. Thus, taking into account Josephson's relation, an expression for the entire electric current through the junction takes the form

$$\frac{h}{2e} C \ddot{\varphi} + \frac{h}{2eR}(1 + \varepsilon \cos \varphi)\dot{\varphi} + I_c \sin \varphi = I_{dc},$$

where I_{dc} is the constant current of the external source, which is considered to be known. After the introduction of dimensionless time $\Omega_d t = \tau$, with $\Omega_d = \frac{2\pi I_c R}{\Phi_0}$ being the Josephson frequency, and $\Phi_0 = h/2e$ being the quantum of magnetic flux, this equation takes the form of (1.1). For dimensionless time, the derivative $\dot{\varphi}$ has a meaning of a voltage with constant component $\langle \dot{\varphi} \rangle_t = \omega$, and parameter γ has a meaning of the constant current through the junction. Thus, in this case, the aforementioned rotation characteristic of rotator has a meaning of a volt-ampere characteristic of a superconductive junction. Consequently, the zero branch of rotation characteristic corresponds to a superconductive branch of volt-ampere characteristic, and rotational branch corresponds to the resistive one. In this case, a complete mechanical and quantum-mechanical analogy occurs.

1.2 Synchronization and Chaotic Rotations of Non-autonomous Rotator

Non-autonomous rotator is governed by a system of the form

$$I\ddot{\varphi} + (1 + F_1(\varphi))\dot{\varphi} + F_2(\varphi) = \gamma + f(\psi),$$
$$\dot{\psi} = \omega_0, \tag{1.8}$$

where $f(\psi)$ is an external excitation with zero mean value $\langle f(\psi) \rangle_\psi = 0$. The explicit time is now "hidden" in the second equation. System (1.8) can be exemplified by: a Froude pendulum, whose shaft performs rotational-oscillatory motion; a superconductive junction placed inside a microwave field; a system of phase synchronization with additive interference, and many other systems. Taking into account a large number of various applications that can be modeled by system (1.8), one can say that this system belongs to the basic models of nonlinear dynamics and theory of oscillations.

The visual simplicity of the model is deceptive. Its analytic investigation is complicated by the complexity of the phase space associated with the existence of monoclinic structures generating chaotic (strange) attractors. For this reason, traditional analytical methods in such a case become mainly ineffective. The analytical results of studies of system (1.8) are not numerous. The majority of them are obtained by qualitative-numerical methods and concern mainly the dynamics of systems of phase synchronization at the boundary of the synchronization capture region. Below, we will attempt to make a feasible contribution to the collection of analytical results concerning rotational motions of the model. Among other things, within the framework of the method of averaging, we will indicate specific ranges of parameters for which rotations of the rotator become chaotic. The investigation is carried out in the quasilinear case $I^{-1} = \mu \ll 1$.

Normally, when a harmonic excitation acts on a quasilinear system, only harmonic resonances at frequencies $n\omega_0$, where n is an integer, are "strong". In terms of an averaging procedure, a resonance is called "strong" if it is "visible" in the averaged system of the first approximation. In this terminology, subharmonic resonances are "weak". It is also known that, among harmonic resonances, the most important is the main (first) resonance $n = 1$, which is called *simple* synchronization.

In other cases, synchronization is called *multiple* synchronization.

1. Let us study a simple forced synchronization of rotations of the rotator. We consider the case of "moderate" amplitudes of the external excitation and solve the problem by transforming system (1.8) into an equivalent system in the standard form:

$$\dot{\xi} = \mu \,\Xi(\xi, \varphi, \psi),$$
$$\dot{\varphi} = \omega_0 + \mu \Phi(\xi, \varphi, \psi),$$
$$\dot{\psi} = \omega_0. \tag{1.9}$$

In contrast to the autonomous case, when the "generating" frequency of the rotator (rotator's frequency for $\mu = 0$) was sought by solving the problem, here it is already determined by the resonance under consideration. The remaining problem is finding functions $\Phi(\xi, \varphi, \psi)$ and $\Xi(\xi, \varphi, \psi)$.

Substituting the second equation of system (1.9) into Eq. (1.8), we obtain:

$$\frac{\partial \Phi}{\partial \varphi}(\omega_0 + \mu \Phi) + \frac{\partial \Phi}{\partial \psi}\omega_0 + \frac{\partial \Phi}{\partial \xi}\dot{\xi}$$
$$+ (1 + F_1(\varphi))(\omega_0 + \mu \Phi) + F_2(\varphi) = \gamma + f(\psi). \tag{1.10}$$

Let us assume that $\gamma - \omega_0 = \mu \Delta$. This condition corresponds to the zone of main resonance by parameter γ (closeness of frequencies). By separating the terms of Eq. (1.10), which are disproportional to μ, we obtain the equation that defines function $\Phi(\xi, \varphi, \psi)$:

$$\frac{\partial \Phi}{\partial \varphi}\omega_0 + \frac{\partial \Phi}{\partial \psi}\omega_0 = -\omega_0 F_1(\varphi) - F_2(\varphi) + f(\psi). \tag{1.11}$$

This equation has a solution of the form

$$\Phi(\varphi, \xi) = -\int F_1(\varphi)d\varphi - \frac{1}{\omega_0}\int F_2(\varphi)d\varphi + \frac{1}{\omega_0}\int f(\psi)d\psi + \xi.$$

The remaining terms of Eq. (1.10) define function $\Xi(\xi, \varphi, \psi)$:

$$\Xi(\xi, \varphi, \psi) = \Delta - \Phi\left(1 + F_1(\varphi) + \frac{\partial\Phi}{\partial\varphi}\right).$$

Thus, functions $\Phi(\xi, \varphi, \psi)$ and $\Xi(\xi, \varphi, \psi)$ are found. After the introduction of a phase mistuning of the form $\eta = \varphi - \psi$, system (1.8) acquires the required standard form of a system with one fast-spinning phase:

$$\dot{\xi} = \mu\,\Xi(\xi, \varphi, \eta),$$
$$\dot{\eta} = \mu\,\Phi(\xi, \varphi, \eta),$$
$$\dot{\varphi} = \omega_0 + \mu\,\Phi(\xi, \varphi, \eta). \qquad (1.12)$$

Let us investigate the dynamics of system (1.12) using specific forms of functions: $F_1(\varphi) = \varepsilon\cos\varphi$, $F_2(\varphi) = \sin\varphi$, $f(\psi) = A\sin\psi$. In this case:

$$\Phi(\varphi, \xi) = -\varepsilon\sin\varphi + \frac{1}{\omega_0}\cos\varphi - \frac{A}{\omega_0}\cos\psi + \xi, \; \psi = \varphi - \eta,$$

$$\Xi(\xi, \varphi, \psi) = \Delta - \left(-\varepsilon\sin\varphi + \frac{1}{\omega_0}\cos\varphi - \frac{A}{\omega_0}\cos\psi + \xi\right)\left(1 - \frac{1}{\omega_0}\sin\varphi\right).$$

By averaging system (1.12), we obtain

$$\dot{\xi} = \mu\left(-\xi + \Delta - \frac{\varepsilon}{2\omega_0} - \frac{A}{2\omega_0^2}\sin\eta\right),$$
$$\dot{\eta} = \mu\xi,$$
$$\dot{\varphi} = \omega_0 + \mu\xi. \qquad (1.13)$$

The first two equations of system (1.13) are independent of the third one and of the variable η, and for a change of time of the form $\mu\sqrt{\frac{A}{2\omega_0^2}}\tau = \tau_n$ they are reduced to the aforementioned equation of the autonomous rotator:

$$\ddot{\eta} + \lambda^r\dot{\eta} + \sin\eta = \gamma^r, \qquad (1.14)$$

where $\lambda^r = \sqrt{\frac{2\omega_0^2}{A}}$, $\gamma^r = \left(\Delta - \frac{\varepsilon}{2\omega_0}\right)\frac{2\omega_0^2}{A}$.

Note that such a situation will be repeated again, which represents a special feature of system transformations that we have performed: namely, if there exists an autonomous system, then in a non-autonomous case, an averaged system will take the form of this autonomous system. The remaining task would be just to properly interpret the properties of the averaged system onto the properties of original system.

Thus, it is possible to use the previously obtained results provided that two questions are answered. The first question is: what is the relation between trajectories of system (1.14) and trajectories of system (1.8)? The second one is: can all bifurcations of Eq. (1.14) be uniquely interpreted onto the corresponding bifurcations of system (1.8)? To answer these questions, we need to refer: to the concept of coarseness of dynamical systems; to the information from the theory of bifurcations regarding a homoclinic trajectory of a saddle and bifurcations of a two-dimensional torus [20, 21], and; to the essence of the averaging procedure as a change of variables [5, 9].

Without going much into detail, it can be argued that if Γ is a trajectory of the averaged system with nonzero characteristic exponents located sufficiently far from the imaginary axis, then this trajectory corresponds to the trajectory $\Gamma \times S^1$ of the original system including peculiarities of its stability. That is, a rough equilibrium of the averaged system will correspond to the limit cycle of the initial system, and a limit cycle of the averaged system will correspond to a two-dimensional torus [8] with a quasi-periodic "winding" (quasi-periodic motions of the system) or with a subharmonic resonance cycle. However, there is a topological equivalence of the phase portrait of the averaged system to the portrait of discrete trajectories in the space of the Poincaré mapping by the secant plane $\varphi(\mathrm{mod}\,2\pi) = \mathrm{const}$ into itself through the period of the fast phase constructed for the original system. Thus, the parameter domains (1), (2), and (3) shown in Fig. 1.1 can be almost completely transferred to the desired bifurcation diagram in the parameter plane $(\gamma^r, \lambda^r) : \lambda \to \lambda^r, \gamma \to \gamma^r$ (see Fig. 1.7a).

The same holds for the phase portraits shown in Fig. 1.1, for which the following designations are to be made: $\varphi \to \eta, \dot{\varphi} \to \xi$, and trajectories shall be considered as trajectories of the point mapping of the form $(\xi, \eta)_\varphi \to \left(\overline{\xi}, \overline{\eta}\right)_{\varphi+2\pi}$ as done for system (1.12), which is equivalent to Eq. (1.8). In this case, equilibria of the initial

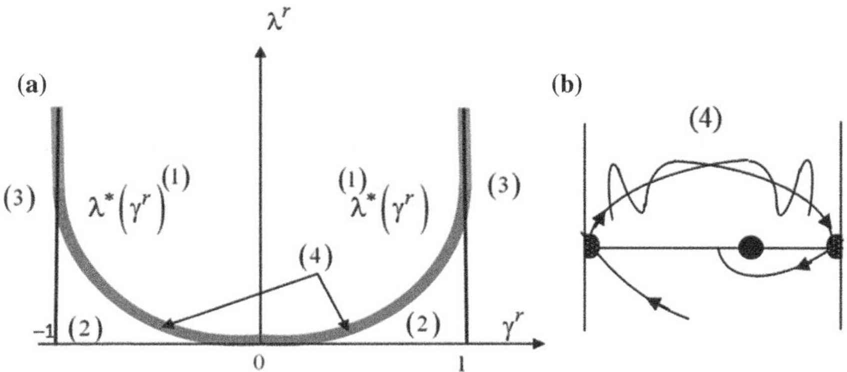

Fig. 1.7 Bifurcation diagram for the parameters that belong to the zone of the main resonance (**a**), and homoclinic structure for the domain (4) (**b**)

system become fixed points of the point mapping, which represent limit cycles of system (1.12), while limit cycles become two-dimensional tori.

Thus, in the parameter domain (1) from Fig. 1.7a, there exist two fixed points: one of them is stable, namely a stable limit cycle corresponding to the synchronization of the rotator by an external excitation, and the second one is a saddle. This parameter domain represents the synchronization acquisition zone, when the synchronism of rotations occurs under any initial condition. In parameter domain (2), in addition to the fixed points mentioned above, there exists an invariant torus. The motions of the system on this torus can be either quasiperiodic (beatings) or periodic (with a slight variation of the parameter) with a period of rotation that has a fractional ratio to the period of external excitation (subharmonic resonance). In such a case, the method of the averaging does not account for such small details. However, since bifurcations on a two-dimensional torus are fully studied, then the aforesaid regarding the fractional resonances can be considered as a statement. The interval of existence of domains (1) and (2), expressed by inequality $|\gamma^r| \leq 1$, i.e. interval of existence of a fixed point of the point mapping, represents the synchronization hold interval. In parameter domain (3), a unique attractor represents a two-dimensional torus. Depending on the system parameters, this torus has either quasiperiodic winding (regime of beatings of rotator), or a limit cycle corresponding to subharmonic synchronization. It is worth mentioning that such resonances are generally weak and disappear during the variation of system parameters. In a general case, when a parameter is varied in domain (3), there is an alternation of regimes of beatings with subharmonic resonances with different ratios of periods to the period of external excitation.

As one can see in Fig. 1.7, instead of the Tricomi curve (corresponding to the separatrix loop of the saddle), there is a region (4) represented as a narrow strip (thick Tricomi curve). It is known that the separatrix loop of a saddle in the phase space is formed from the confluence ("sticking together") of the incoming and outgoing manifolds—separatrices of the saddle. This is the only possible behavior for phase trajectories. Unlike phase trajectories, the intersection of trajectories of a point mapping is not forbidden. In this case, the intersection of the incoming and outgoing manifolds of the saddle (Fig. 1.7b) represents a general case. On the other hand, the intersection that has arisen once entails an infinite number of intersection points. This is so because the intersection point belongs to both manifolds simultaneously, and, consequently, it must belong to both manifolds in further iterations of the point mapping. That is, each intersection point is a pre-image of another point, and the latter is also a point of intersection of invariant manifolds. The incoming and outgoing manifolds oscillate, intersecting each other over an infinite number of points with an oscillation step tending to zero as it approaches the saddle point from the point of intersection. Points of intersection of manifolds are called the homoclinic points. The set of homoclinic points forms two sequences: one sequence tends to a saddle for $t \to +\infty$, while the other tends to the saddle for $t \to -\infty$. The set of homoclinic points represents a trace of a phase trajectory, which is double asymptotic to a saddle: a homoclinic Poincaré curve. It is proven that, in a vicinity of homoclinic curves, there is a countable set of saddle limit cycles [21], which have their own homoclinic curves, and so on. The homoclinic structure, as an assembly of homoclinic curves and an infinite set of saddle limit cycles in the case under consideration, represents an attracting trajectory for neighboring trajectories and, therefore, is a chaotic attractor. It must be said that in such structures there could be stable limit cycles, but they have so small areas of attraction that the natural noises that are always present in the real system hinder their realization. In other words, under real conditions, homoclinic structures are "real" strange attractors. The second mechanism for the formation of dynamic chaos in the system under consideration is a bifurcation of the destruction of a two-dimensional torus [20, 21] that exists in parameter domain (3) as parameters approach domain (4). In the autonomous case, for the parameters that belong to the Tricomi curve, the limit cycle was "stuck" in the separatrix loop of the saddle. But for the torus, this is impossible: the torus first loses its smoothness and then collapses with the further formation of a chaotic attractor in its place. When the parameters are varied within parameter domain (4), the aforementioned attractors merge.

Thus, parameter domain (4) corresponds to the chaotic dynamics of the rotator. This domain is rather narrow: its width is $\sim\mu$ and it shrinks to the Tricomi curve as the small parameter decreases. Note that the existence of separatrix loops of saddles in the averaged system is almost a guarantee of the existence of homoclinic structures in the initial system and that could play the role of a criterion.

Let us construct a part of the rotation characteristic of a non-autonomous rotator in the zone of the main resonance, having noted that, according to the definition of the rotation characteristic and from the principle of averaging, it follows that $\langle \dot{\varphi}^* \rangle_t = \omega_0 + \langle \dot{\eta}^* \rangle_t$, while from the designation of parameters it follows that $\gamma^r \sim \Delta \sim \gamma$. Consider $\lambda^r = \mathrm{const}$, for which the corresponding line on Fig. 1.7 crosses domains (1), (4), (2), and (3) when parameter γ^r is increased from zero. A complete picture of the rotation characteristic is obtained using the central symmetry of this curve (in the first approximation) in the zone under consideration.

Suppose that parameter γ^r belongs to domain (1). In the regime of synchronization of the rotor's rotations, $\langle \dot{\eta}^* \rangle_t = 0$, and, therefore, $\langle \dot{\varphi}^* \rangle_t = \omega_0$. The rotor "holds" in the synchronization regime until a stable fixed point on the line $\gamma^r = 1$ does not disappear. Therefore, the rotation characteristic has a vertical section at the frequency of external excitation $\Omega = \omega_0$. This vertical section represents a resonance step defined by inequality $|\gamma^r| \leq 1$, or, in extended form: $\omega_0 - \mu \frac{A}{2\omega_0^2} \leq \gamma \leq \omega_0 + \mu \frac{A}{2\omega_0^2}$. The center of this step belongs to the rotational branch of the rotation characteristic of an autonomous rotator (see Fig. 1.8).

For $|\gamma^r| = 1+0$ [i.e. during the transition of parameters from domain (2) to domain (3)], a limit cycle disappears and the rotator sharply transitions from the regime of synchronization into the regime of quasiperiodic beatings that corresponds to torus T^2. In this case, $\langle \dot{\eta}^* \rangle_t \neq 0$, and $\langle \dot{\eta}^* \rangle_t$ increases as γ is increased and, consequently, decreases when γ is decreased. In both cases, a sequence of subharmonic resonances may appear on the torus, which is reflected on rotation characteristic in the form of small vertical steps (they are not shown in Fig. 1.8). When parameter γ enters domain (4), the torus loses its smoothness, is destroyed, and a strange attractor is formed in its place [20]. Next, the homoclinic structure of the saddle fixed point of the main resonance is mixed with this attractor. A burst of intensity of chaotic rotations

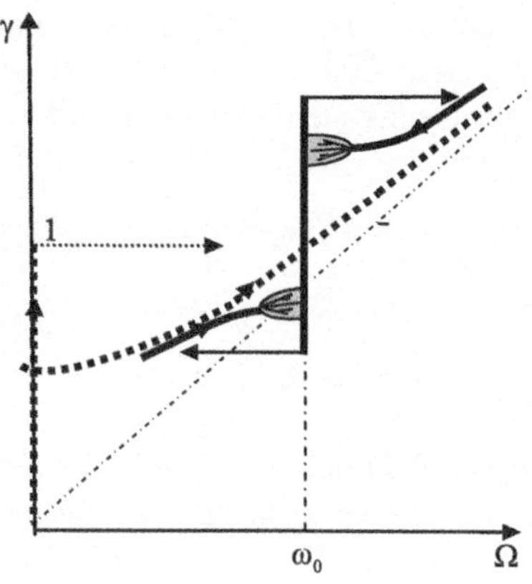

Fig. 1.8 Rotation characteristic of non-autonomous rotator (——) in the zone of main resonance

takes place. With a further decrease of the parameter, the strange attractor loses its attraction property and the rotator sharply returns to the resonance step. On the left-hand side of the resonance stage, the dynamics of the rotator undergoes the same bifurcations and the behavior of the rotation characteristic is the same. The rotation characteristic in the regime of synchronization is similar to that of the autonomous rotator except in regions determined by strange attractors.

When system trajectories belong to a chaotic attractor, the structure of the rotation characteristic is complicated. Details of its behavior in such cases will be discussed at the end of the section.

1. Now let us consider the dynamics of rotator at large amplitudes of the external excitation $A \sim \mu^{-1} \gg 1$. In this case, we will be interested in the dynamics of system in the zone of arbitrary harmonic resonance. This case is technically more complicated than the previous one. To reduce Eq. (1.8) to the standard form, a preliminary transformation is required.

For simplicity, we will assume $\varepsilon = 0$ and make a change of the variables: $\varphi = b\sin(\psi + \psi_0) + \theta$, $\psi + \psi_0 = \psi_n$ (further, subscript n will be omitted), $\psi_0 = \mathrm{arctg}\frac{1}{\omega_0 I}$, $b = \dfrac{A}{\omega_0\sqrt{\omega_0^2 I^2 + 1}} \approx \dfrac{A}{\omega_0^2 I}$. As a result, (1.8) reduces to an equivalent equation of the form

$$I\ddot{\theta} + \dot{\theta} + \sin(\theta + b\sin\psi) = \gamma. \qquad (1.15)$$

It is known that

$$\sin(\theta + b\sin\psi) = J_0\sin\theta + \sum_{m=1}^{\infty} J_{2m}\{\sin(2m\psi + \theta) - \sin(2m\psi - \theta)\}$$

$$+ \sum_{k=0}^{\infty} J_{2k+1}\{\sin((2k+1)\psi + \theta) + \sin((2k+1)\psi - \theta)\},$$

where $J_l = J_l(b)$ are the Bessel functions of the first kind with an integer argument. Now transforming Eq. (1.15) according to the already known rule, we obtain an equivalent system in the standard form

$$\dot{\xi} = \mu^{1/2}(\gamma - n\omega_0 \pm J_n(b)\sin\eta) - \mu\xi - \mu\xi\frac{\partial\Phi}{\partial\theta} - \mu^{3/2}\left(1 + \frac{\partial\Phi}{\partial\theta}\right),$$

$$\dot{\theta} = n\omega_0 + \mu^{1/2}\xi + \mu\Phi(\theta, \psi),$$

$$\dot{\psi} = \omega_0, \tag{1.16}$$

where notation «+» is used for even numbers n, and «−» for odd numbers. In this case, an equation that defines function Φ has the form

$$n\omega_0\frac{\partial\Phi}{\partial\theta} + \omega_0\frac{\partial\Phi}{\partial\psi} = -J_0\sin\theta$$
$$- \sum_{m=1}^{\infty} J_{2m}\{\sin(2m\psi + \theta) - \delta_{n,2m}\sin(2m\psi - \theta)\} \tag{1.17}$$
$$- \sum_{k=0}^{\infty} J_{2k+1}\{\sin((2k+1)\psi + \theta) + \delta_{n,2k+1}\sin((2k+1)\psi - \theta)\},$$

where $\delta_{n,l} = 1$, for $n \neq l$, and $\delta_{n,l} = 0$ for $n = l$.

Finding the solution of Eq. (1.17) is not difficult. By introducing a phase mistuning $\eta = \theta - n\psi$ and averaging system (1.16) up to terms of the order of $\sim\mu$, we obtain the following averaged system:

$$\dot{\xi} = \mu^{1/2}(\gamma - n\omega_0 \pm J_n(b)\sin\eta) - \mu\xi,$$
$$\dot{\eta} = \mu^{1/2}\xi,$$
$$\dot{\theta} = n\omega_0 + \mu^{1/2}\xi.$$

Implementing a time change $\sqrt{\mu|J_n(b)|}\tau = \tau_n$, we reduce the first two equations of this system to the already known equation:

$$\ddot{\eta} + \lambda_n^r\dot{\eta} \pm \sin\eta = \gamma_n^r, \tag{1.18}$$

where $\lambda_n^r = \sqrt{\dfrac{\mu}{|J_n(b)|}}$, $\quad \gamma_n^r = \dfrac{\gamma - n\omega_0}{|J_n(b)|}$.

Note that signs in front of the function are not of fundamental importance. If necessary, one could remove the sign «−» by applying transformation $\eta \to \eta + \pi$. There is an interesting fact regarding Eq. (1.18): if one would forget that A in this equation is a "large" parameter, then the argument of the Bessel function is sufficiently small and $J_1\left(\frac{A}{\omega_0^2 I}\right) \approx \frac{A}{2\omega_0^2 I}$. In this case, Eq. (1.18) turns into Eq. (1.14). This can be considered as another "miracle of a small parameter", when a result that has been obtained under certain conditions remains valid for other conditions.

Let us analyze Eq. (1.18). In a general case, for large amplitudes of the external excitation, all the harmonic resonances of the quasilinear rotator are essential. The magnitude of the resonance steps equals $2|J_n(b)|$. Without taking into account strange attractors [that belong to parameter domain (4)], the qualitative picture of rotation characteristic in the vicinity of each of these resonances is similar to the rotation characteristic of the autonomous rotator (with or without a hysteretic loop). Neighboring resonance steps on the rotation characteristic can overlap. The overlap value of adjacent resonances $\delta\gamma$ is defined by inequality $(n+1)\omega_0 - |J_{n+1}(b)| < \delta\gamma < n\omega_0 + |J_n(b)|$. In this interval, one or another resonance is realized depending on the initial conditions of the system. From this inequality and from the properties of the Bessel functions [22], it follows that there is no overlap of the resonances for $\omega_0 > 1, 2$, and if ω_0 is sufficiently small, then the overlap occurs not only for the neighboring resonances. Figure 1.9 exemplifies graphs of several Bessel functions.

It follows from the properties of these functions that hysteretic properties of the rotation characteristic strongly appearing for small n disappear as soon as the number of resonance is increased. A transition to the non-hysteretic rotation characteristics may be non-monotonous. Moreover, for the values of amplitudes that correspond to the roots of equation $J_n(b) = 0$, resonance with number n is just absent. For example, for the main resonance for $A = \nu\omega_0^2 I$, $\nu \approx 3.9; \; 7.0; \; 10.2; \; \dots$. For the second resonance, $\nu \approx 5.2; \; 8.4; \; \dots$. Thus, by changing the parameter b, we can obtain a long gallery of different rotation characteristics. Figure 1.10 shows a qualitative example of the rotation characteristic of non-autonomous rotator for high amplitudes of the external excitation (subharmonic resonances are not shown).

Within the framework of the method of averaging, we have obtained a general picture of dynamical processes. Such an a priori knowledge reduces further experimental or numerical investigation of system dynamics into the constructive work. Without knowledge of the overall picture, the researcher could be considered as a "blind kitten" drowning in a sea of seemingly disconnected facts. In particular, using

Fig. 1.9 Bessel functions

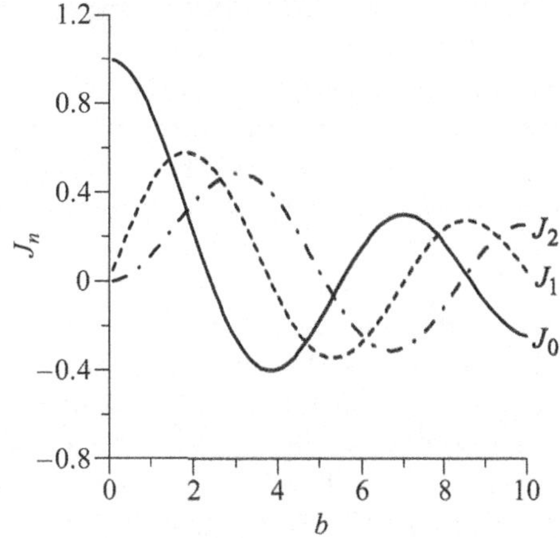

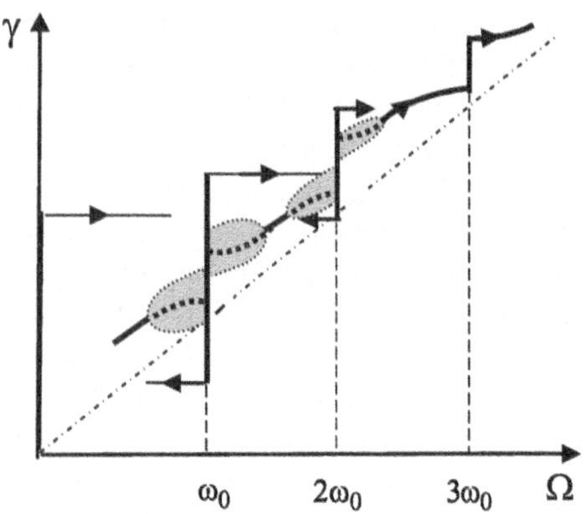

Fig. 1.10 Rotation characteristic of non-autonomous rotator, a "devil's" stairway of resonances. Subharmonic resonances are not shown here

the obtained information as an introductory one, a numerical investigation can then consist of clarifying the details of processes themselves and their bifurcations; refining the boundaries of the necessary dynamic regimes in the parameter space, and; extrapolating the results onto non-small values of parameters, etc.

A numerical study has been carried out for functions $F_1(\varphi) \equiv 0$, $F_2(\varphi) = \sin \varphi$, $f(\psi) = A \sin \psi$ for the following values of parameters $\mu = 0.2$, $A = 3.5$, $\omega_0 = 0.75$. Parameter γ has been varied in the interval [1.1177; 1.8248], covering two resonance zones. For simplicity, system (1.8) was written in the form

$$\dot{\varphi} = \gamma + x,$$
$$\dot{x} = \mu(-x - \sin \varphi + A \sin \psi),$$
$$\dot{\psi} = \omega_0. \tag{1.19}$$

Figure 1.11 shows the gallery of projections of trajectories of system (1.19) onto the involute of phase cylinder (bold lines) as well as trajectories of Poincaré mapping $(\varphi, \xi)_{\psi=\psi_0} \to \left(\overline{\varphi}, \overline{\xi}\right)_{\psi=\psi_0+2\pi}$ constructed for this system (dotted lines).

Starting with $\gamma = 1.8248$, we start to decrease this parameter; physically, it corresponds to the decrease in torque of the pendulum. For this parameter value, the pendulum performs quasi-periodic rotations, whose image is an ergodic torus. This dynamical regime is realized for arbitrary initial conditions (it is globally stable). Figure 1.11a shows quasi-periodic winding of this torus, while Fig. 1.11b shows invariant curve of the Poincaré mapping (compare to Fig. 1.1d). During the decrease of parameter γ, we enter a grey zone (the one to the right of the second step shown in Fig. 1.10), and a subharmonic resonance with rotation number of 11/5 arises (stable and unstable limit cycles, see Fig. 1.11c). The value of a corresponding resonance step is very small (for this reason, such resonances cannot be "caught" using the method of the averaging). Further, limit cycles move closer to each other and disappear. A new torus is then formed. In the space of the Poincaré mapping, the scenario of merging of limit cycles is almost the same as a scenario of merging of equilibria with further origination of a limit cycle from the separatrix loop of saddle-node in the

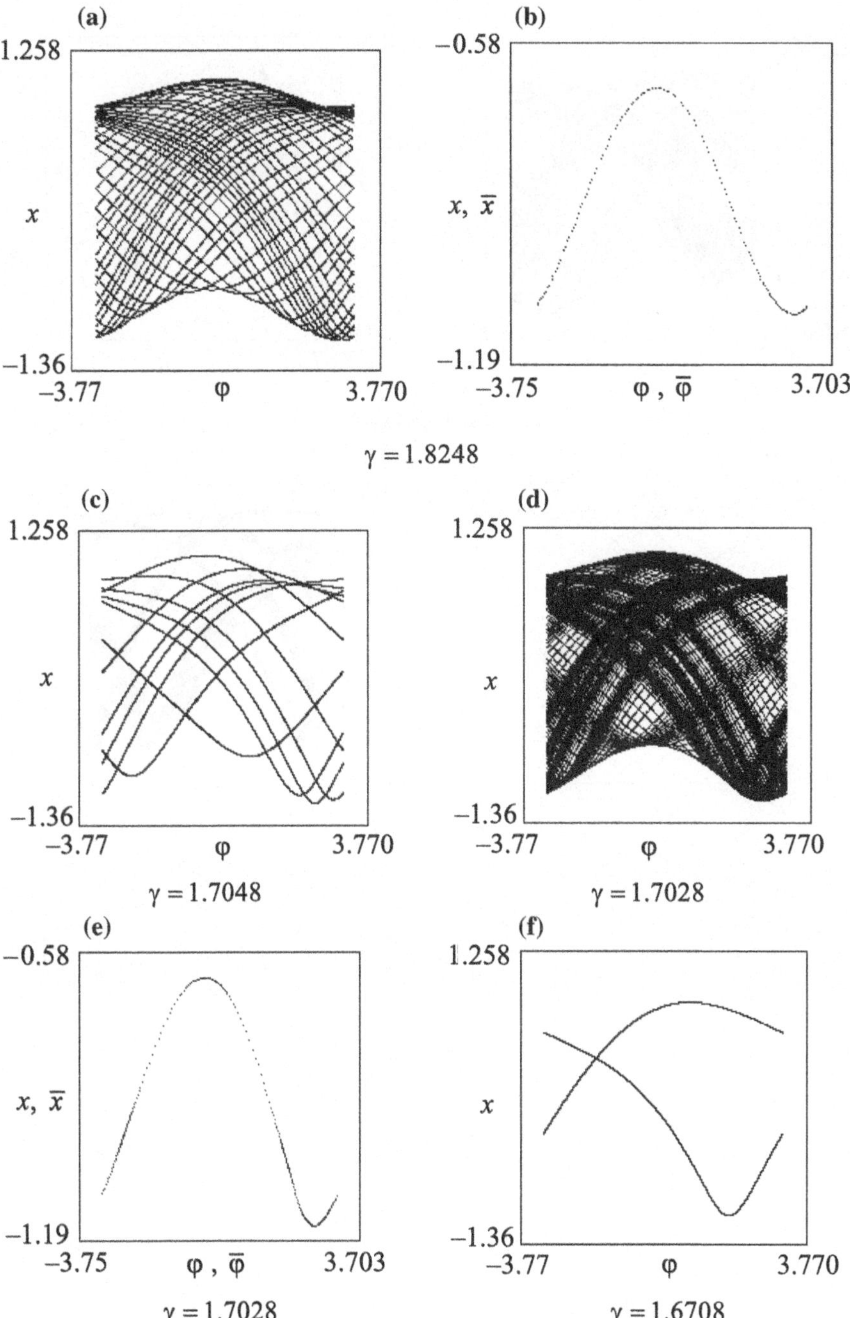

Fig. 1.11 The gallery of qualitatively different phase portraits and Poincare mappings for the different γ varied from 1.8248 down to 1.1177

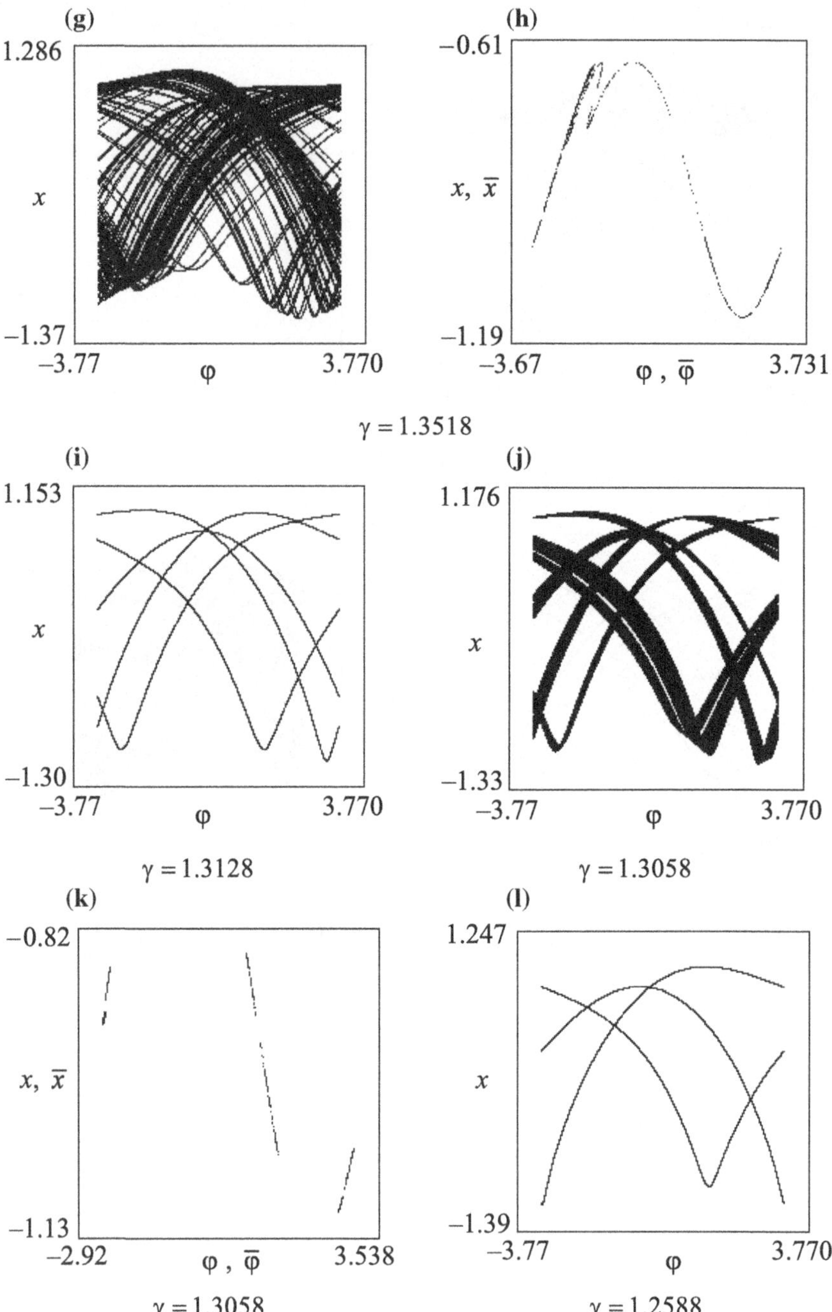

Fig. 1.11 (continued)

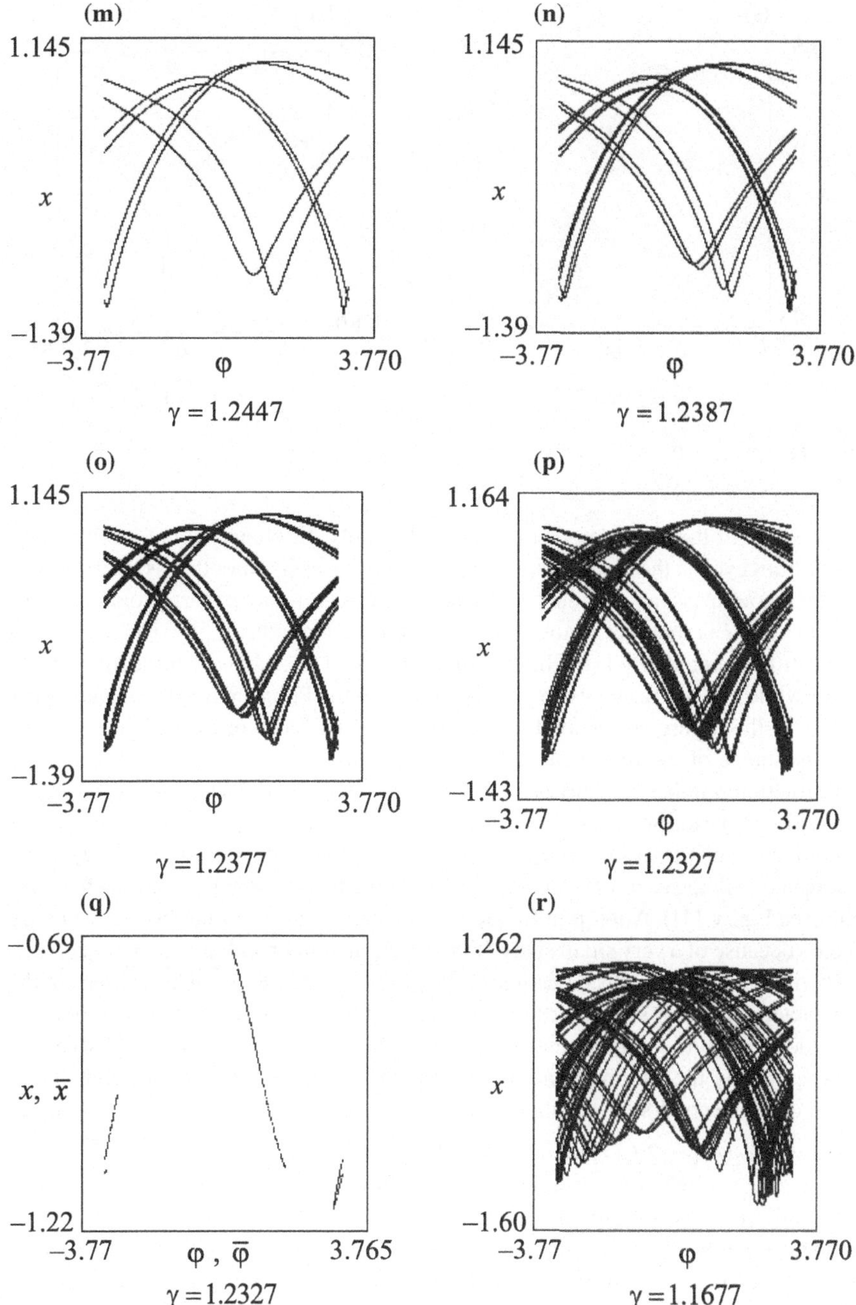

Fig. 1.11 (continued)

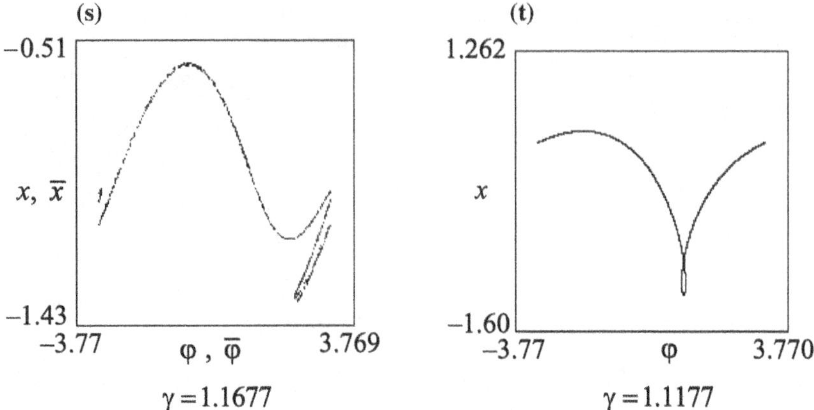

Fig. 1.11 (continued)

phase space of the autonomous pendulum. The only difference is: before the merging of limit cycles, the "old" torus (as a manifold) loses its smoothness. During such loss of smoothness, an origination of a chaotic attractor may occur. In our case, this effect is so weak that it is almost invisible. At least, the interval of existence of such attractor by parameter γ is negligibly small. Figure 1.11d shows a projection of this torus, while Fig. 1.11e shows an invariant curve of the Poincaré mapping. Dark areas shown in this figure correspond to the thickening of trajectories, which represents an inheritance of the disappeared limit cycle. Further, a new fractional resonance with rotation number 15/7 arises at the appeared torus and disappears according to the same scenario. Such scenario is repeated over and over until the rotator jumps from one of such fractional resonances (at the moment of its disappearance) to the resonance with the rotation number of 2, i.e. to the regime of harmonic synchronization (see Fig. 1.11f). A jump in frequency of rotation from sub-harmonic resonance occurs because of a very small size of the attraction domain of limit set of trajectories that originates after the disappearance of the limit cycle. Very small deviations of the parameter (in fourth sign after the comma) lead to the loss in system stability and its sudden jump to the regime of harmonic synchronization. "Fast-living" dynamical regimes by the parameter are not of a physical interest. In our experiment, the value of the second step equals $\delta_2(\gamma) = 0.32$. Just to compare: $b = \dfrac{A}{\omega_0\sqrt{\omega_0^2 I^2 + 1}}$,
$\delta_2(\gamma) = 2|J_2(b)| = 2J_2(1.2024) = 0.3198$.

 For $\gamma = 1.35$, the rotator suddenly jumps from the resonance step to the regime of chaotic rotations. On the graph of rotation characteristic (see Fig. 1.10), such a jump occurs towards the left-hand part of the grey zone adjacent to the second step. At this time, the chaotic attractor is already formed. Figure 1.11g shows a small time interval of trajectories, since otherwise trajectories would not be distinguishable and the whole square area of a graph would be completely black.

At the Poincaré plane, there is a torus with a "fold" and "appendices" (see Fig. 1.11h) that indicates a chaotic character of the motion. To clarify types of bifurcations that lead to chaos, parameter γ has been increased with a smaller step. It has been found that the chaotic attractor originates according to the following scenario: "subharmonic resonance at torus $\rightarrow$ torus loses smoothness $\rightarrow$ doubling of the period of stable resonance limit cycle $\rightarrow$ strange attractor" [20]. Sometimes this attractor is called the chaotic torus-attractor [23].

With further decrease of parameter γ, a stable limit cycle (subharmonic resonance with rotation number 5/3) originates at a chaotic attractor (internal bifurcation, see Fig. 1.11i). This resonance experiences a cascade of period doubling with further formation of the Feigenbaum attractor. Figure 1.11p shows a small interval of phase trajectories, while Fig. 1.11q shows Poincaré mapping. There is an ambiguous branch in the right bottom corner. Further, a chaotic attractor related to homoclinic trajectories of saddle resonance limit cycle (with rotation number 1/1) is mixed with this attractor. A "splash" in the chaotic state of rotator occurs (see Figs. 1.11r, s). Further, the frequency of rotation of the rotator jumps to the value $\Omega = \omega_0 = 0.75$, which means a jump to the regime of simple synchronization. Figure 1.11t shows a first resonance with a jump for $\gamma = 1.1177$.

Scattering of rotation characteristic. When the rotator moves at chaotic attractor (see Fig. 1.11r, s), its average velocity $\langle \dot{\varphi}^*(t, t_0) \rangle_t = \omega_0 + \langle \dot{\eta}^*(t, t_0) \rangle_t$ strongly depends on the initial conditions, i.e. on those phase coordinates for which the rotator enters the regime of chaotic rotations which, in principle, is determined by the entire dynamic history of the rotator. This can be hardly controlled in any numerical experiment. Therefore, when repeating the same experiment, each time the passage through the regime of chaotic rotations will take place over different trajectories that have different mean values. As a result, one obtains an ambiguous rotation characteristic with an infinite number of branches, which has the form of a "broom". In this case, going back to the synchronization regime (sudden jump) will occur for different values of parameter γ. The effect will be "amplified" by the finiteness of the time when averaging and due to the presence of natural noises. Such an effect was observed for the first time during the study of dynamics of systems with superconductive Josephson junctions and it was regularly observed in real experiments [24].

References

1. Andronov, A.A., Vitt, A.A., Khaikin, S.E.: Theory of Oscillators. Dover Publications, New York (1987)
2. Barbashin, E.A., Tabueva, V.A.: Dynamic Systems with a Cylindrical Phase Space. Nauka, Moscow (1969). (in Russian)
3. Bautin, N.N., Leontovich, E.A.: Methods and Means for a Qualitative Investigation of Dynamical Systems on the Plane. Moscow (1976) (In Russian)
4. Van der Pol, B.: Forced oscillation in circuit with non-linear resistance. Philos. Mag. Ser. 7 **3**(13), 65–80 (1927)

5. Krylov, N.M., Bogolyubov, N.N.: Introduction to Non-linear Mechanics. Princeton University Press (1947) (Translated from Russian)
6. Hale, J.K.: Oscillations in nonlinear systems. McGrawHill, New York (1963)
7. Mitropolsky, Y.A., Lykova, O.V.: Integral Manifolds in Nonlinear Mechanics. Nauka, Moscow (1973). (in Russian)
8. Neimark, Y.I.: The Method of Point Mappings in the Theory of Non-linear Oscillations. Moscow (1972) (In Russian)
9. Volosov, V.M., Morgunov, B.I.: Averaging Methods in the Theory of Non-linear Oscillatory Systems. Moscow (1971) (In Russian)
10. Likharev, K.K.: Dynamics of Josephson Junctions and Circuits. Gordon and Breach, New York (1986)
11. Andronov, A.A., Leontovich, E.A., Gordon, I.I., Maier, A.G.: Theory of bifurcations of dynamical systems on a plane. Israel Prog. Sci. Trans. (1973)
12. Magnus, K., Schwingungen. Stuttgart: Teubner, (1976)
13. Gorev, A.A.: Selected Works on Questions of the Stability of Electric Systems. Gosenergoizdat, Moscow (1960). (in Russian)
14. Korolev, V.I., Fufaev, N.A., Chesnokova, R.A.: Correctness of approximate study of rotor vibration in a synchronous machine. Appl. Mathem. Mech. **6**(37), 1007–1014 (1973). (in Russian)
15. Akimov, N., Belyustina, L.N., Belykh, V.N., et al.: Phase Synchronization Systems. Radio Svyaz', Moscow) (1982). (Russian)
16. Lindsey, W.C.: Synchronization Systems in Communication Control. Prentice-Hall, Inc., New Jersey EnglewoodCliffs (1972)
17. Josephson, B.D.: The discovery of tunnel supercurrents. Rev. Mod. Phys. **46**, 252–254 (1974)
18. Barone, A., Paterno, G.: Physics and Applications of the Josephson Effect. Wiley, New York (1982)
19. McCumber, D.E.: Effect of AC impedance on DC voltage-current characteristics of superconductor weak-link junctions. J. Appl. Phys. **39**, 3113 (1968)
20. Afraimovich, V.S., Shilnikov, L.P.: Invariant tori, their breakdown and stochastisity. Amer. Math. Soc. Transl. **149**, 201–211 (1991)
21. Arnold, V.I. (ed.), Gamkrelidze, R.V. (ed.-in-chief): Progress in Science and Technology. Current Problems in Mathematics. Fundamental Directions 3 (Dynamical Systems III), VINITI AN SSSR, Moscow (1985)
22. Abramowitz, M., Stegun, I.A.: Handbook of mathematical functions. Dover Publications Inc., New York (1964)
23. Anishchenko, V.S.: Dynamical chaos basic concepts. Teubner-Texte zur Physik, Leipzig (1987)
24. Belykh, V.N., Pedersen, N.F., Soerensen, O.H.: Shunted Josephson junction model: part 1 autonomous case, part 2 nonautonomous case. Phys. Rev. B. **16**(11), 4853–4871 (1977)

Chapter 2
Autonomous and Non-autonomous Systems with One and a Half Degrees of Freedom

Intensive vibrations of unbalanced rotors do not only shorten the lifetime of expensive equipment, but can also lead to catastrophic failures. Impressive evidence of such accidents has been collected in [1]. In addition to the natural unbalance of the rotors, incorrect engineering solutions can also lead to undesired vibrations. In particular, it may happen that the embedded additional vibration dampers, instead of the positive effect (and contrary to intuition), lead to an increase in the instability of rotors.

Generally speaking, stability or instability of systems is an area of formal knowledge and intuition may prove to be a bad adviser in dealing with problems related to stability of systems. For this reason, there is no other way than to study the dynamics of models starting with simplest ones and continuing by gradually increasing the number of degrees of freedom.

2.1 Dynamics of Rotator with Aperiodic Load

Figure 2.1 shows a cross-section of the Froude pendulum, whose sleeve has an additional internal sleeve. There is viscous damping in between the sleeves just as in between the shaft and the external sleeve. The shaft rotates with a constant angular speed ω. $\delta_1(\Omega - \dot{\varphi})$ is the moment of the friction force acting out of the shaft and towards the pendulum. $\delta_2(\omega - \dot{\varphi})$, is the moment of the friction force acting from the internal sleeve, $mgl \sin\varphi$ is the moment of the gravity force, $\delta_3(\Omega - \omega)$, and $\delta_2(\omega - \dot{\varphi})$, are moments of the friction forces acting from the shaft towards the internal sleeve and acting from the external sleeve, respectively. $\delta_1, \delta_2, \delta_3$ are the constant parameters. The governing equations have the form

$$I_1\ddot{\varphi} = \delta_1(\Omega - \dot{\varphi}) + \delta_2(\omega - \dot{\varphi}) - mgl \sin\varphi,$$
$$I_2\dot{\omega} = \delta_3(\Omega - \omega) - \delta_2(\omega - \dot{\varphi}),$$

© Springer Nature Switzerland AG 2020

N. Verichev et al., *Chaos, Synchronization and Structures in Dynamics of Systems with Cylindrical Phase Space*, Understanding Complex Systems,
https://doi.org/10.1007/978-3-030-36103-7_2

Fig. 2.1 Cross-section of the Froude pendulum with extra internal sleeve. *ES* is the external sleeve, *IS* is the internal sleeve, and *S* is the shaft

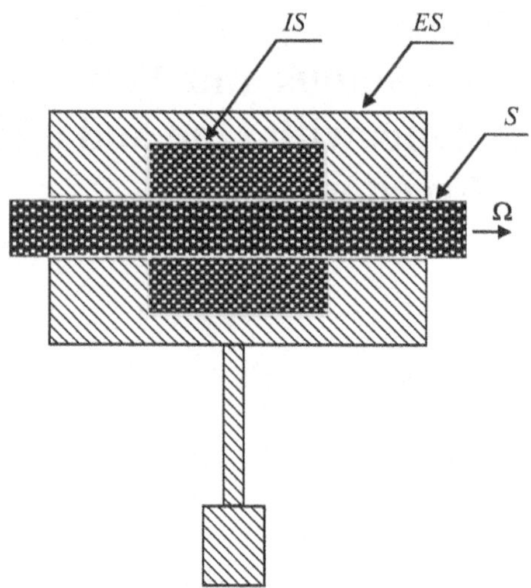

where I_1, I_2 are the normalized moment of inertia of the pendulum and moment of inertia of the internal sleeve, respectively.

Using the transformation of variable $J = \frac{\delta_2}{mgl}(\omega - \dot{\varphi}) - \frac{\delta_2\delta_3}{mgl(\delta_2+\delta_3)}\Omega$ and time $\frac{mgl}{\delta_1}\tau = \tau_n$, we reduce these equations to the form

$$I\ddot{\varphi} + \dot{\varphi} + \sin\varphi = \gamma + J,$$
$$\dot{J} + hJ = a\dot{\varphi} + b(\sin\varphi - \gamma), \qquad (2.1)$$

where $I = \frac{I_1 mgl}{\delta_1^2}$, $\gamma = \frac{\Omega}{mgl}\left(\frac{\delta_1\delta_2+\delta_1\delta_3+\delta_2\delta_3}{\delta_2+\delta_3}\right)$, $h = \frac{\delta_1}{mgl}\left(\frac{\delta_1+\delta_2}{I_2} + \frac{\delta_2}{I_1}\right)$, $a = \frac{\delta_2}{mgl}\left(\frac{\delta_1}{I_1} - \frac{\delta_3}{I_2}\right)$, $b = \frac{\delta_1\delta_2}{I_1 mgl}$.

Consider one more system: an asynchronous electric motor of finite power is coupled to an elastic shaft with a rigidly coupled unbalanced disk (see Fig. 2.2). An inertial sleeve is placed over the shaft. There is viscous damping in between the shaft and the sleeve. The latter cannot move along the shaft.

Stabilizing properties of the inertial sleeve are quite straightforward: when the operational speed of rotation fluctuates towards higher values, the sleeve starts consuming energy of the motor, while for an opposite fluctuation, a part of kinetic energy of the sleeve is transmitted to the shaft. The balancing sleeve plays a role of an integrator (averaging component), which is analogous to properties of a frequency filter in electric engineering.

Let us assume that torque is a linear function of the form $M_d = M_0 - \lambda\dot{\varphi}$ with $M_0 = \text{const}$, and $\lambda\dot{\varphi}$ is the normalized moment of friction forces. Moments that act on the rotor: $mg\varepsilon\sin\varphi$ is the moment of the gravity force of the unbalanced mass m that has eccentricity ε; $\delta(\omega - \dot{\varphi})$ is the moment of forces of viscous damping acting from the sleeve; $-\delta(\omega - \dot{\varphi})$ is the moment that acts on the sleeve from the shaft. In this case, the governing equations take the form

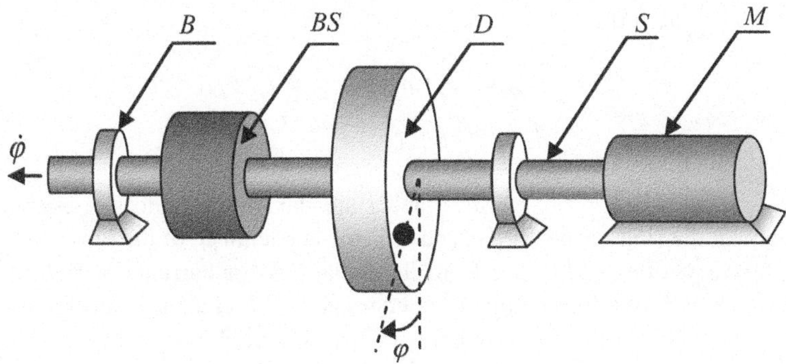

Fig. 2.2 Asynchronous electric motor with unbalanced disk and balancing sleeve. M is the motor, D is the disk, BS is the balancing sleeve, S is the shaft, and B is the bearing

$$I_1\ddot{\varphi} = M_0 - \lambda\dot{\varphi} + \delta(\omega - \dot{\varphi}) - mg\varepsilon\sin\varphi,$$
$$I_2\dot{\omega} = -\delta(\omega - \dot{\varphi}), \tag{2.2}$$

where I_1, I_2 are the normalized moment of inertia of the pendulum and moment of inertia of the internal sleeve, respectively; m is the imbalance; ε is the eccentricity; λ, and δ are the coefficients of viscous damping.

Using the transformation of variable $\frac{\delta}{mg\varepsilon}\omega = J$ and time $\frac{mg\varepsilon}{\lambda+\delta}\tau = \tau_n$, Eqs. (2.2) are reduced to system (2.1) with the following parameters: $I = \frac{I_1 mg\varepsilon}{(\lambda+\delta)^2}$, $\gamma = \frac{M_0}{mg\varepsilon}$, $h = \frac{\delta(\lambda+\delta)}{I_2 mg\varepsilon}$, $a = \frac{\delta^2}{I_2 mg\varepsilon}$.

One more example: Fig. 2.3a shows a scheme of a superconductive quantum interferometer (R-SQUID), which is used for the measurement of extra-small magnetic fields [2]. Its equivalent electric scheme is shown in Fig. 2.3b.

We obtain a dynamical model of this system using Kirchhoff laws and a resistive model of the junction. For dimensionless variables and parameters, the governing

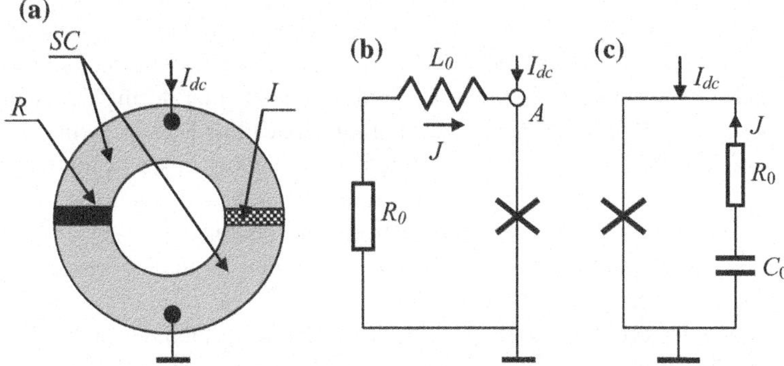

Fig. 2.3 **a** a scheme of the R-SQUID, where R is the resistor, SC are the superconductors, I is the isolator, **b** equivalent electric scheme, **c** superconductive junction loaded by RC-chain

equations have the form:

$$C\ddot{\varphi} + \dot{\varphi} + \sin\varphi = \gamma + J,$$

$$\dot{J} + hJ = a\dot{\varphi},$$

where $\frac{I_{dc}}{I_c} = \gamma$, $\frac{r}{l} = h$, $-\frac{1}{l} = a$; r and l are the dimensionless resistance and inductance of the ring, respectively; and C is the capacitance of the junction.

The first equation is obtained from the equations for currents at node A. The second one comes from the equality of voltages at the RL-chain and superconductive junction. One can see that the dynamics of the system shown in Fig. 2.3c is also governed by system (2.1). Because of the fact that the equations of the aforementioned systems are described by the same system of equations, accurate within designations and physical sense of the variables, this system deserves special attention.

Let us study dynamical properties of system (2.1). This system has two equilibria $O(\varphi_0, \dot{\varphi}_0, J_0)$: $O_1(\arcsin\gamma, 0, 0)$ and $O_2(\pi - \arcsin\gamma, 0, 0)$. Equilibrium O_1 is stable for $|\gamma| < 1$ and represents either a node or a focus, depending on the parameters of the system. Equilibrium O_2, depending on the parameters, can either represent a saddle-node or a saddle-focus. $\dim W^s = 2$, $\dim W^u = 1$, where W^s, W^u are the stable and the unstable manifolds of equilibrium O_2, respectively.

Oscillatory phase trajectories (bounded by φ) do not exist in the phase space $G(\varphi, \dot{\varphi}, J)$. In particular, if $Ib(h - a) + b - a > 0$ (which is valid for the systems under consideration), then this fact can be proven with the help of the periodic Lyapunov function of the form:

$$V = \frac{Ib(h - a) + b - a}{2} I\dot{\varphi}^2 + \frac{1}{2}J^2 + bI\dot{\varphi}J$$

$$+ \left(b + bhI - a - b^2 I\right) \int_{\varphi_0}^{\varphi} (\sin\varphi - \gamma)d\varphi.$$

Its derivative, taken along the trajectories of system (2.1) $\dot{V} = -(Ib(h - a) + b - a)\dot{\varphi}^2 - (h - b)J^2 \leq 0$, is negative in the whole phase space.

Thus, stationary oscillatory motions (oscillatory limit cycles) cannot exist in the system under consideration. In practical terms, it means that during the change of initial conditions, the Froude pendulum and unbalanced disk in the course of time reach either equilibrium or a rotary dynamical regime. Consequently, the Josephson junction in the course of time enters either a superconductive state or a generation regime. Rotary limit cycles can originate from the separatrix loop of a saddle via bifurcation of a doubled limit cycle that can appear due to the thickening of phase trajectories.

Let us consider structures of rotary trajectories in more detail. Excluding variable J and using the transformation of time, we reduce system (2.1) to the equivalent equation of the third order

$$\beta\,\dddot{\varphi} + \ddot{\varphi} + \alpha\cos\varphi + \lambda\dot{\varphi} + \sin\varphi = \gamma, \tag{2.3}$$

where $\beta = \sqrt{\dfrac{I^2 h}{(1+hI)^3}}$, $\alpha = \dfrac{1}{\sqrt{h(1+hI)}}$, $\lambda = \dfrac{h-a}{\sqrt{h(1+hI)}}$.

Equation (2.3) is known in the theory of synchronization [3]. Its properties crucially depend on ratio α/β. It is not difficult to see that $\alpha > \beta$. In this scenario, there exist two cases. In the first one, the dynamics of the system is regular for all values of parameters. The bifurcation diagram on the plane (γ, λ), $\alpha = $ const, $\beta = $ const has a form shown in Fig. 2.4. In the second case, there exists a parameter domain on the bifurcation diagram for which the system has chaotic dynamics, see Fig. 2.5.

Consider the first case. The curve $\lambda = \lambda^0(\gamma)$ on the bifurcation diagram corresponds to the doubled limit cycle that has originated due to increased density of the

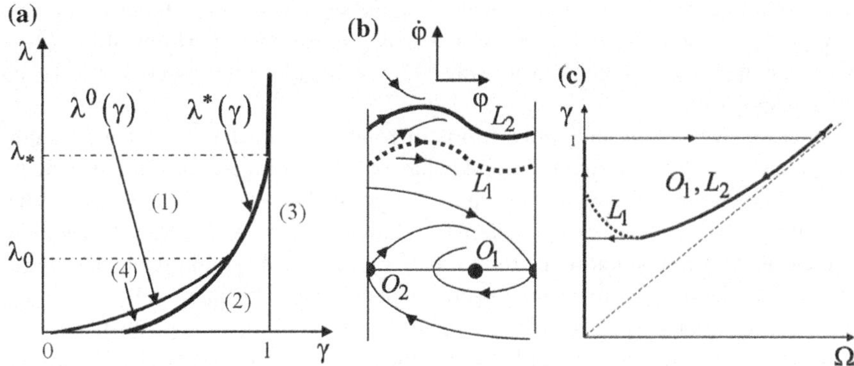

Fig. 2.4 **a** Bifurcation diagram in the case of regular dynamics, **b** two dimensional analogue of the phase portrait for the parameters that belong to domain (4), **c** qualitative form of the rotation characteristic for $\lambda < \lambda_0$

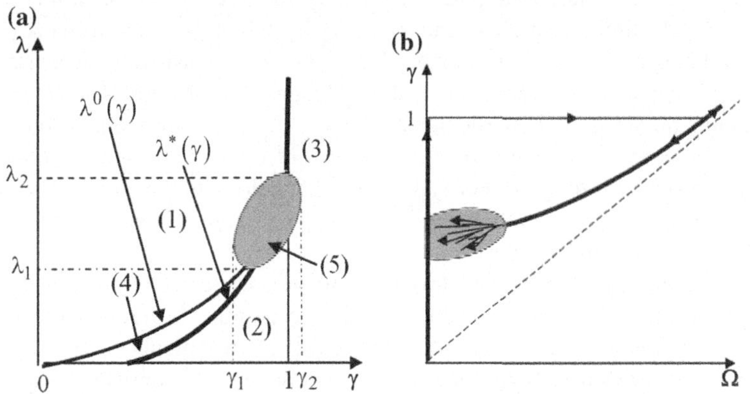

Fig. 2.5 Bifurcation diagram in the case of existence of dynamical chaos (**a**), rotation characteristic including effect of scattering for $\lambda_1 < \lambda < \lambda_2$ (**b**)

phase trajectories. At the moment of origination, the cycle has a positive frequency of rotation $\Omega > 0$. When parameters are varied out of parameter domain (1) and into domain (4), a doubled limit cycle is decomposed into two limit cycles: stable and unstable ones. A two-dimensional analogue of the phase portrait of Eq. (2.3) at the involute of the phase cylinder is shown in Fig. 2.4b. For increased γ, the stable limit cycle moves upwards over the phase cylinder, while the unstable one moves downwards. For the parameters that belong to the Tricomi curve, $\lambda = \lambda^*(\gamma)$ and $\lambda < \lambda_0$, the unstable limit cycle gets stuck in the separatrix loop of saddle O_2 and then disappears. For the opposite variation of γ, the stable limit cycle moves downwards. For the parameters that belong to the Tricomi curve, an unstable limit cycle originates from the separatrix loop and then moves towards the stable limit cycle. On the curve $\lambda = \lambda^0(\gamma)$, these limit cycles stick together and disappear. In the parameter domain (1), equilibrium O_1 is globally stable. For $\lambda_0 < \lambda < \lambda_*$, a stable limit cycle with zero rotational frequency originates from the separatrix loop, and its frequency increases as γ increases. Two-dimensional analogues of phase portraits of Eq. (2.3) for the parameters that belong to domains (1), (2), and (3) are shown in Fig. 1.1b, c, and d, respectively.

As usual, we obtain qualitative forms of rotation characteristics of the rotator using phase portraits, bifurcations, and behavior of limit trajectories of the system. Let us vary parameter γ along the line $\lambda = $ const. For any $\lambda = $ const from the interval $\lambda > \lambda_*$, the rotation characteristic qualitatively coincides with the rotation characteristic of the autonomous rotator (see Fig. 2.4a). A qualitative matching of rotation characteristics also occurs for any λ that belongs to the interval $\lambda_0 < \lambda < \lambda_*$ (see Fig. 2.4b). For $\lambda < \lambda_0$, we obtain a qualitative form of the rotation characteristic as shown in Fig. 2.4c. Let us remark that parameter γ is varied quasi-statically starting with smaller values if it belongs to parameter domain (1). In this case, for any initial conditions, the rotator arrives at equilibrium O_1 (see Fig. 1.1a) and stays there until this equilibrium disappears for $\gamma = 1+0$, without "feeling" bifurcations and changes of phase portraits. This governs the "zero" branch of the rotation characteristic. For $\gamma = 1+0$, the rotator jumps into the rotational regime (see Fig. 1.1d). Further on, the frequency of the limit cycle grows and the rotary branch of the rotation characteristic grows and gets closer to the asymptote (bisector). For a quasi-static decrease of γ, the rotator "stays" on the limit cycle without "feeling" the changes that occur in the surrounding phase space. In particular, it does not "feel" that for γ that belong to the Tricomi curve, an unstable limit cycle, a "killer" of rotations originates and moves towards the stable limit cycle. For values of γ that belong to the curve $\lambda^0(\gamma)$, these limit cycles approach each other and disappear. The rotator jumps back to the zero branch of the rotation characteristic.

Consider the second case. In contrast to the previous case, the bifurcation diagram has an extra "grey" domain (5) (see Fig. 2.5a), corresponding to the chaotic dynamics of the system. Let us assume that parameters γ, λ are varied along the Tricomi curve toward the higher values. Then, for $\lambda < \lambda_1$, a unique, unstable limit cycle originates from the separatrix loop O_2 (parameters are varied from domain (2) to domain (4)). In this case, it has a so-called saddle value $\sigma_s > 0$ [4]. For $\lambda > \lambda_2$, a stable limit cycle gets stuck in the separatrix loop (transition of parameters from domain (2) to domain (1)). In this case, $\sigma_s < 0$, i.e. a change of the sign of the saddle value takes place. Apart from that, for the trajectory motion along a two-dimensional stable manifold W^s of equilibrium O_2, a bifurcation of origination of focus from the saddle occurs and at that moment the saddle-focus value $\sigma_{sf} > 0$. Thus, one can conclude that there exist parameters for which conditions of the Shilnikov theorem (about the existence of a countable set of periodic motions of the saddle type near a loop of a saddle-focus) are satisfied (see [4] for more details). It is worth mentioning that, apart from the loop of the saddle-focus, by varying parameters inside the parameter domain (5), one could observe other bifurcations, such as a chain of doublings of a period of limit cycle with further origination of the Feigenbaum attractor as well as merging of chaotic attractors ("crisis of attractors" or "alternation of chaoses"). We will pay attention to all details during the discussion of results of the numerical experiment.

Numerical experiment. Let us discuss the dynamics of the system during the variation of parameters inside the grey zone. By decreasing the constant component of the torque of the rotator (parameter γ) and by keeping all other system parameters constant: $\mu = 0.08$, $h = 0.05524$, $a = -0.0809$, $b = 0.021$. For these parameters, $\lambda = 0.567$.

Figure 2.6 shows phase portraits and portraits of the point mapping $(x, J)_{\varphi=\varphi_0} \rightarrow (\bar{x}, \bar{J})_{\varphi=\varphi_0+2\pi}$ obtained for an equivalent dynamical system of the form

$$\dot{\varphi} = \gamma + x,$$
$$\dot{x} = \mu(-x - \sin\varphi + J),$$
$$\dot{J} = -hJ + a(\gamma + x) + b(\sin\varphi - \gamma).$$

For $\gamma = 1.29$ (and for $\gamma > 1.29$), the rotator performs periodic motions (see Fig. 2.6a). For $\gamma = 1.282$, the limit cycle experiences the first bifurcation of doubling of the period (pitchfork bifurcation), see Fig. 2.6b. Then a cascade of pitchfork bifurcations occurs, see Figs. 2.6c-f, that ends by origination of the Feigenbaum attractor. Figure 2.6h shows the curve of a point mapping that corresponds to this strange attractor. The form of this curve suggests that, by using a proper choice of coordinates, this mapping can be reduced to the classical "parabola mapping". With the decrease of γ, a stable rotary-oscillatory limit cycle of a tripled period (with the main period equal to 2π) is born on the chaotic attractor. A typical feature of this limit cycle consists of a rotation of the phase at a two-dimensional stable manifold. The motion at the limit cycle occurs in such a way that the aforementioned equilibrium would exist in the phase space of the system (saddle-focus with two-dimensional stable and one-dimensional unstable manifolds).

One can say that this limit cycle is an inheritance of the Shilnikov attractor. For the decreased γ, this limit cycle loses its stability and a stable limit cycle of the period five is born instead. Such a scenario repeats over and over, which eventually leads to the chaotic motion of the rotator at the united chaotic Shilnikov-Feigenbaum attractor. During the merging of attractors, a splash of values of the chaotic motion (maximum Lyapunov exponent, dimension, pedestal of spectrum, etc.) with respect

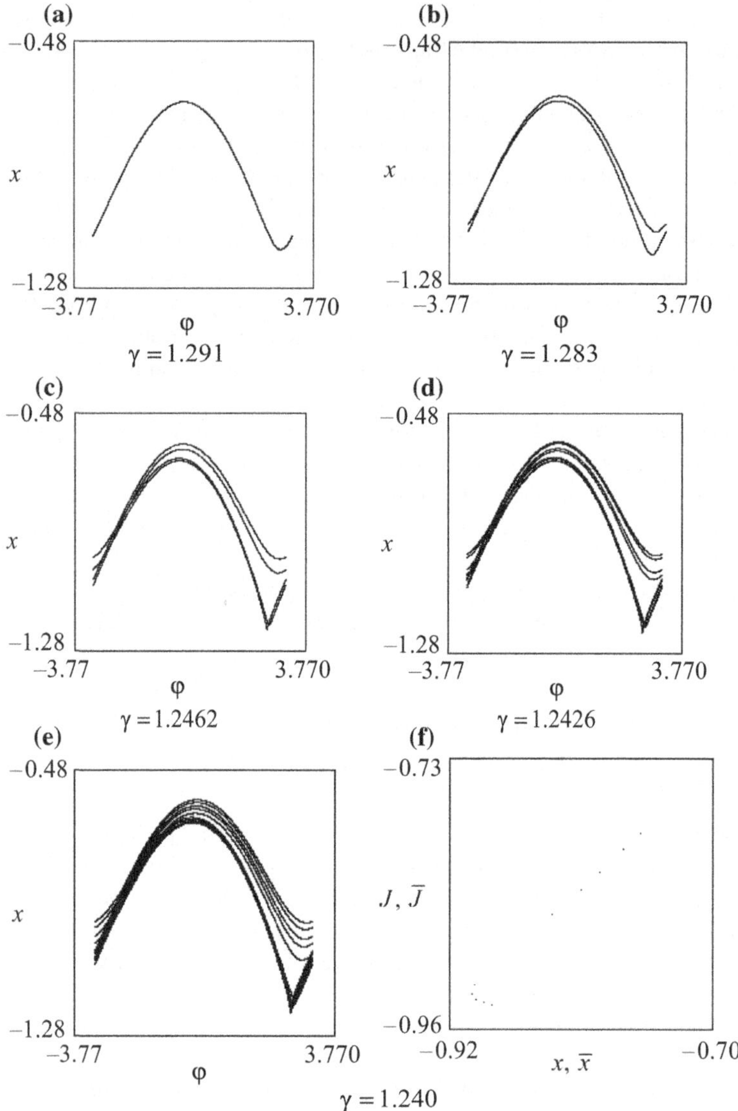

Fig. 2.6 The gallery of qualitatively different phase portraits and Poincare mappings for the different γ varied from 1.291 down to 1.091 (continues on the next pages)

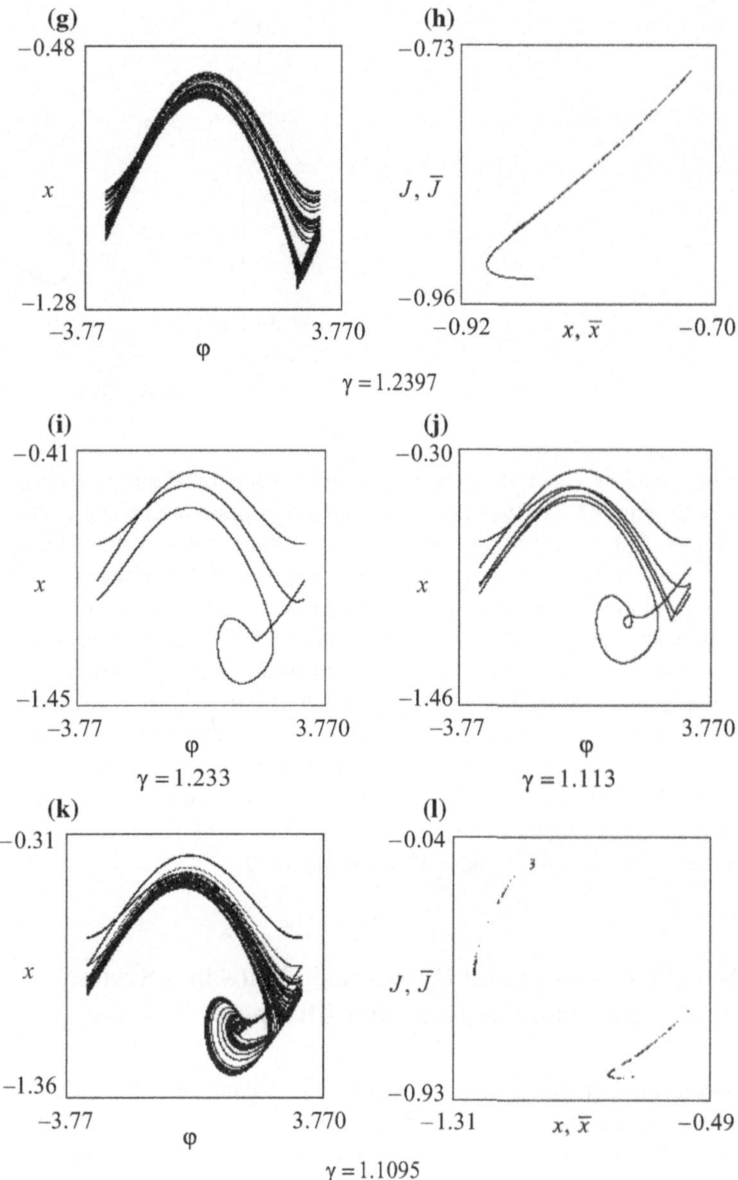

$$\gamma = 1.2397$$

$$\gamma = 1.233 \qquad\qquad \gamma = 1.113$$

$$\gamma = 1.1095$$

Fig. 2.6 (continued)

Fig. 2.6 (continued)

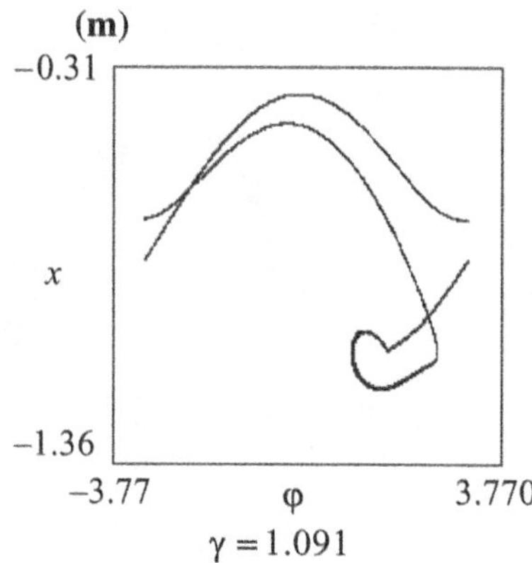

to the corresponding characteristics of motion on a separated attractor occurs. The Poincaré mapping that corresponds to the motion at the united chaotic attractor is shown in Fig. 2.6l. With further decrease of γ, a stable limit cycle of the doubled period is born, see Fig. 2.6m, and the rotator jumps from the rotational regime to equilibrium. Of course, these scenarios take place for the parameters used in this numerical experiment. For other parameters, degeneration of chaos occurs through the bifurcation of the separatrix loop of saddle-focus. Normally, it is quite challenging to accurately follow bifurcation scenarios since the corresponding parameter domain is very narrow. The aforementioned mechanism of chaotization of the dynamical processes is well known in the dynamics of quantum interferometers [2].

The considered system represents just one of the examples of how the original dynamics of a simple system may be complicated after the introduction of devices that seem to be "harmless", such as a balancing sleeve.

2.2 Synchronization and Dynamical Chaos in a System of Non-autonomous Rotator with Aperiodic Load

Let us again consider the Froude pendulum with a balancing sleeve (see Fig. 2.1a). We will assume that the shaft of the pendulum rotates with the frequency $\Omega(t) = \Omega_0 + \Omega_s \sin \psi$, $\dot{\psi} = \varpi_0$, In this case, the governing equations have the form

$$I_1 \ddot{\varphi} = \delta_1(\Omega_0 - \dot{\varphi}) + \delta_2(\omega - \dot{\varphi}) - mgl \sin \varphi + \delta_1 \Omega_s \sin \psi,$$

$$I_2 \dot{\omega} = \delta_3(\Omega_0 - \omega) - \delta_2(\omega - \dot{\varphi}) + \delta_3 \Omega_s \sin \psi,$$

$$\dot{\psi} = \varpi_0. \tag{2.4}$$

Let us perform a series of transformations of system (2.4) to reduce it to the form which is analogous to the case of the autonomous rotator. First of all, let us remove

the periodic function from the second equation by using a change of the variables of
the form $\omega = A_s \sin\psi + A_c \cos\psi + y$,

where $A_s = \dfrac{\delta_3(\delta_2+\delta_3)}{(I_2\varpi_0)^2+(\delta_2+\delta_3)^2}\Omega_s$, $A_c = -\dfrac{I_2\delta_3\varpi_0}{(I_2\varpi_0)^2+(\delta_2+\delta_3)^2}\Omega_s$.

As a result, we obtain a system of the form

$$I_1\ddot\varphi + \delta_1\dot\varphi + mgl\,\sin\,\varphi = \delta_1\Omega_0 + \delta_2(y-\dot\varphi)y + B\,\sin(\psi+\psi_0),$$
$$I_2(y-\dot\varphi)^{\cdot} + (\delta_2+\delta_3)(y-\dot\varphi) = \delta_3\Omega_0 - \delta_3\dot\varphi - I_2\ddot\varphi,$$
$$\dot\psi = \varpi_0 \tag{2.5}$$

where $B = \sqrt{(\delta_2 A_s + \delta_1\Omega_s)^2 + (\delta_2 A_c)^2}$, $\psi_0 = \text{arctg}\dfrac{\delta_2 A_c}{\delta_2 A_s + \delta_1\Omega_s}$.

Further, similarly to the autonomous case, we transform time in (2.5) $\frac{mgl}{\delta_1}\tau = \tau_n$
and variables $J = \frac{\delta_2}{mgl}(y-\dot\varphi) - \frac{\delta_2\delta_3}{mgl(\delta_2+\delta_3)}\Omega_0$, and $\psi + \psi_0 = \psi_n$. We obtain the
following system:

$$I\ddot\varphi + \dot\varphi + \sin\varphi = \gamma + J + A\sin\psi,$$
$$\dot J + hJ = a\dot\varphi + b(\sin\varphi - \gamma - A\sin\psi),$$
$$\dot\psi = \omega_0, \tag{2.6}$$

where $A = \frac{B}{mgl}$, $\omega_0 = \frac{\delta_1}{mgl}\varpi_0$, and all other parameters are the same as ones in the
autonomous case. Subscript "n" is omitted here.

Consider dynamics of system (2.6) for the different asymptotic values of parameter
I.

<u>Asymptotic case $I \ll 1$</u>. In this case, according to [5, 6], a singularly per-
turbed system (2.6) has a stable integral surface of "slow motions" in the phase
space $G(\varphi, \psi, \dot\varphi, J)$. Integral curves at this surface represent trajectories of the
non-autonomous rotator. Its parameters are close ($\sim I$) to those that are obtained
from system (2.6) for $I = 0$. As a result of transformations, the "generating"
system is reduced to the equation of non-autonomous rotator (1.8) with functions
$F_1(\varphi) = \varepsilon\cos\varphi$, $F_2(\varphi) = \sin\varphi$. In this case, the combinations of parameters of
system (2.6), $\frac{h-b}{(h-a)^2}$, $\frac{1}{(h-a)}$, γ, $\frac{\sqrt{\omega_0^2+(h-b)^2}}{h-b}A$, $\frac{h-a}{h-b}\omega_0$ play the role of parameters
$I, \varepsilon, \gamma, A, \omega_0$ in Eq. (1.8), respectively. This case is already discussed in detail in
Sect. 1.2.

<u>Asymptotic case $I^{-1} = \mu \ll 1$</u>. In this case, we need to apply additional trans-
formations to reduce system (2.6) to the standard form (see Appendix I for details).
Consider the following parameter domain:

$$D_\mu = \left\{ I^{-1} = \mu,\ h = \mu h^*, a = \mu a^*, b = \mu b^*, \frac{h-a}{h-b}\left(\frac{h-b}{h-a}\gamma - \omega_0\right) = \mu\Delta \right\}.$$

Physically, this domain corresponds to the zone of a simple master-slave
synchronization of rotator with aperiodic load that has low damping.

We apply to (2.6) a change of the variables of the form

$$\dot{\varphi} = \omega_0 + \mu \Phi(\varphi, \psi, \xi),$$

$$J = \frac{a-b}{h-b}\omega_0 + \mu Y(\varphi, \psi) + \mu\, y. \tag{2.7}$$

Equations for functions $\Phi(\varphi, \psi, \xi)$, $Y(\varphi, \psi)$ and their solutions, as well as the constant of the second equation, can be found using the same methodology as described in Sects. 1.2 and 2.1.

Inside domain D_μ, the change of the variables (2.7) reduces system (2.6) to the standard form:

$$\dot{\xi} = \mu\left(y + Y - \Phi\frac{\partial\Phi}{\partial\varphi} - \Phi + \Delta\right),$$

$$\dot{y} = \mu\left(-h^*y - h^*Y - \Phi\frac{\partial Y}{\partial\varphi} + a^*\Phi - b^*\Delta\right),$$

$$\dot{\eta} = \mu\Phi(\xi, \varphi, \eta),$$

$$\dot{\varphi} = \omega_0 + \mu\Phi(\xi, \varphi, \eta), \tag{2.8}$$

where $\Phi = \frac{1}{\omega_0}\cos\varphi - \frac{A}{\omega_0}\cos\psi + \xi$, $Y = \frac{b^*}{\omega_0}\cos\varphi + \frac{Ab^*}{\omega_0}\cos\psi$, $\eta = \varphi - \psi$ is the phase mistuning, and Δ is the frequency mistuning.

Having averaged system (2.8) over a fast-spinning phase, we obtain a truncated system of the form

$$\dot{\xi} = \mu\left(-\xi + y - \frac{A}{2\omega_0^2}\sin\eta + \Delta\right),$$

$$\dot{y} = \mu\left(-h^*y + a^*\xi + b^*\left(\frac{A}{2\omega_0^2}\sin\eta - \Delta\right)\right),$$

$$\dot{\eta} = \mu\xi,$$

$$\dot{\varphi} = \omega_0 + \mu\xi. \tag{2.9}$$

In system (2.9), we have kept old notations for the averaged variables. By transforming time $\mu\frac{2\omega_0^2}{A}\tau = \tau_n$ and variable $y = \frac{A}{2\omega_0^2}x$, we reduce the first three equations of system (2.9) to the equivalent system of the form

$$I_r\ddot{\eta} + \dot{\eta} + \sin\eta = \gamma^r + x,$$

$$\dot{x} + h_r x = a_r\dot{\eta} + b_r\left(\sin\eta - \gamma^r\right), \tag{2.10}$$

where $I_r = \frac{A}{2\omega_0^2}$, $\gamma^r = \Delta I_r^{-1}$, $h_r = h^*I_r^{-1}$, $a_r = a^*I_r^{-1}$, $b_r = b^*I_r^{-1}$.

It is easy to see that the obtained averaged system (2.10), accurate within notations of variables and parameters, coincides with system (2.6). Parameters of equivalent equation of the third order (2.3) are related to those of system (2.6) in the following way:

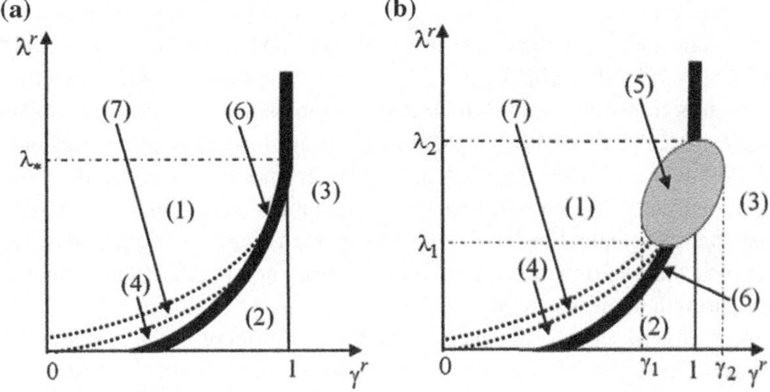

Fig. 2.7 Bifurcation diagrams of parameters of system (2.10) in a zone of the main resonance

$$\beta = \sqrt{\frac{A}{2\omega_0^2}\frac{(h^* - b^*)}{(1 + h^*)^3}},\, \alpha = \sqrt{\frac{A}{2\omega_0^2}\frac{1}{(h^* - b^*)(1 + h^*)}},$$

$$\lambda^r = \sqrt{\frac{2\omega_0^2}{A}}\frac{h^* - a^*}{\sqrt{(h^* - b^*)(1 + h^*)}},\, \gamma^r = \left(\frac{A}{2\omega_0^2}\right)^{-1}\Delta.$$

Figure 2.7 shows bifurcation diagrams of the parameters of system (2.10) as analogues of corresponding diagrams for the autonomous system shown in Figs. 2.4 and 2.5. In both figures, parameter domain (1) corresponds to the existence of a saddle limit cycle and a stable limit cycle. The latter is globally stable and corresponds to the synchronization of the rotators' rotations. This domain is called the synchronization capture domain, which means that for arbitrary initial conditions the rotator reaches the synchronization regime. Parameter domain (2) corresponds to the existence of a saddle limit cycle and a stable limit cycle, as well as a stable invariant torus T^2. For the parameters that belong to this domain, depending on the initial conditions, either a regime of synchronization or a regime of stable stationary beatings can occur. Domain (3) corresponds to the existence of a globally stable invariant torus. For the parameters that belong to this domain, a regime of simple synchronization does not take place. Note that in all cases we consider tori as invariant manifolds, not mentioning any structures of trajectories on these tori. Their manifolds can have either quasi-periodic "winding" or limit cycles that correspond to subharmonic resonances. If, for certain conditions, a torus is realized in the system, then in one case this torus may correspond to the regime of quasi-periodic beatings, while for a small variation of parameters it may correspond to the regime of subharmonic synchronization. However, we can state that the smaller the parameter μ is, the less the probability of existence of subharmonic resonance is. For parameter domain (4), there exist stable and saddle limit cycles and stable and unstable tori. In contrast to Figs. 2.4 and 2.5, Fig. 2.7 shows the existence of two new domains (domains (6) and (7)). Domain (6) ("bold Tricomi curve") corresponds to chaotic attractors that are related to the destruction of a two-dimensional invariant torus and also to the existence of rough homoclinic curves of a saddle resonance limit cycle. Most likely, domain (7) ("bold curve of limit cycle of doubled period") also corresponds to chaotic attractors. For the parameters transition from domain (4) to domain (1), a merging of stable and unstable tori occurs. Until nowadays, a bifurcation of merging of tori has not been studied. Probably, before the merging, tori lose their smoothness. Normally, this leads to the origination of chaotic attractors. The aforementioned domains are "grey" in the sense that their exact borders are not determined and exact bifurcations leading to the chaotic state of dynamical processes are not known. All of the details can be clarified during a numerical study.

Let us discuss the correspondence of chaotic attractors of initial system (2.8) and averaged system (2.10). It is clear that if A is a strange attractor of the averaged system, then the attractor of the original system does not represent a formal multiplication $A \times S^1$ due to non-roughness of attractor A (homoclinic trajectories do not have an unambiguous interpretation). On the other hand, with the decrease of parameter μ, domains (6) and (7) shown in Fig. 2.7 tighten towards the analogous bifurcation curves shown in Figs. 2.4 and 2.5, while domain (5) shrinks to the corresponding domain shown in Fig. 2.5. The corresponding bifurcation curves in these domains are getting closer, respectively. In particular, it takes place for curves that correspond to the doubling of the period of limit cycle in system (2.4) and for curves corresponding to doubling of one of the periods of torus in system (2.6). In other words, one can state that for any $\mu = \mu_0$, a chaotic attractor of the original system contains $N(\mu_0)$ number of tori of saddle type (a "skeleton" of the chaotic attractor). Each of these tori corresponds to a limit cycle of the averaged system and $\lim_{\mu_0 \to 0} N(\mu_0) = \infty$.

Due to this reason, hereinafter, we will state that a chaotic attractor has the same type as a chaotic attractor of the averaged system, keeping in mind that this state-ment is not entirely correct. In this sense, bifurcations of tori in system (2.6) should take place according to the same scenario as bifurcations of limit cycles in aver-aged system (2.10). Phase portraits of system (2.1) and forms of the trajectories of point mappings for system (2.6) (on the corresponding planes) should have identical qualitative features.

Note that Eq. (2.10) are invariant with respect to the transformation $(\eta, x, \gamma^r) \to (-\eta, -x, -\gamma^r)$. This means that bifurcation diagrams shown in Fig. 2.7 shall also contain their mirrored reflection with respect to the axis λ^r. As a result, one will obtain a full picture of bifurcations of dynamical regimes that occur during the approach to the resonance step from the left-hand side as from the right-hand one. That is, if parameter γ^r moves from parameter domain (3) into parameter domain (1) (see Fig. 2.7), then it corresponds to the motion towards resonance from the right-hand side. For a similar motion of parameter γ^r in the left-hand side of the diagram, it corresponds to the motion towards resonance from the left-hand side.

Numerical study. This study is focused on bifurcations of dynamical regimes of the rotator during the transition from the regime of beatings to the regime of synchronization, which corresponds to the transition of parameter γ (while all other parameters remain fixed) from domain (3) to domain (1), see Fig. 2.7.

When parameter γ is varied through domain (6), the character of bifurcations is the same as that for the non-autonomous rotator described in Sect. 1.2, therefore, this case can be ignored. Let us discuss the case of parameter transition from domain (3) to domain (1) through domain (5), since this is the most interesting case in terms of the demonstration of theoretical results.

In this numerical study, phase portraits and structures of point mapping $(\varphi,\, x,\, J)_{\psi=\psi_0} \rightarrow \left(\overline{\varphi}, \overline{x}, \overline{J}\right)_{\psi=\psi_0+2\pi}$ have been analyzed. This point mapping has been made for a system which is equivalent to system (2.6):

$$\dot{\varphi} = \gamma + x,$$
$$\dot{x} = \mu(-x - \sin\varphi + J + A\sin\psi),$$
$$\dot{J} = -hJ + a(\gamma + x) + b(\sin\varphi - \gamma - A\sin\psi),$$
$$\dot{\psi} = \omega_0.$$

Experiment 1. The following parameter set has been used: $\mu = h = 0.006$, $a = -0.006$, $b = 0.003$, $A = 5$, $\omega_0 = 0.5$. For these parameters, $\lambda^r = 0.632$. The gallery of phase portraits and their corresponding point mappings are shown in Fig. 2.8. Parameter γ is varied towards the resonance step from the right-hand side.

For $\gamma = 2.0892$, the rotator experiences quasi-periodic beatings, see Fig. 2.8a. The left figure shows a torus with quasi-periodic winding. The middle and right figures represent a trajectory of point mapping of this torus. Discretization of the point mapping trajectory is not explicit due to the high density of the points. For the decreased γ, a bifurcation of doubling of the period of torus (pitchfork bifurcation) occurs, see Fig. 2.8b. Further, the torus changes its configuration with no further pitchfork bifurcations, see Fig. 2.8c. Then an opposite pitchfork bifurcation occurs, see Fig. 2.8d. The obtained torus loses its stability and the rotator suddenly jumps into the synchronization regime, see Fig. 2.8e. This figure shows a trajectory of the limit cycle at the involute of the phase cylinder.

For $\gamma = 1.9412$, the rotator suddenly jumps from the synchronization regime into the regime of quasi-periodic beatings (a jump from the lower end of the resonance step), see the torus of the doubled period shown in Fig. 2.8f (compare to Fig. 2.8c). If parameter γ would be increased, then the torus would experience an opposite pitchfork bifurcation similar to the transition from Fig. 2.8c, d. Further, a loss of the stability of this torus and a jump towards the resonance step from the left-hand side would occur. A slight visual difference of cases shown in Fig. 2.8c, f is concerned with the choice of a secant plane. For the transition from Fig. 2.8f, h, the torus changes its configuration and then experiences an opposite pitchfork bifurcation, see Fig. 2.8i.

Thus, in this case, dynamics of the rotator during the transition to the synchronization regime is regular. Note that bifurcation scenarios during the transition to the regime of synchronization from the left-hand and from the right-hand sides to the resonance step are the same. This has been declared above as a consequence of the invariance of the averaged system to the applied transformations.

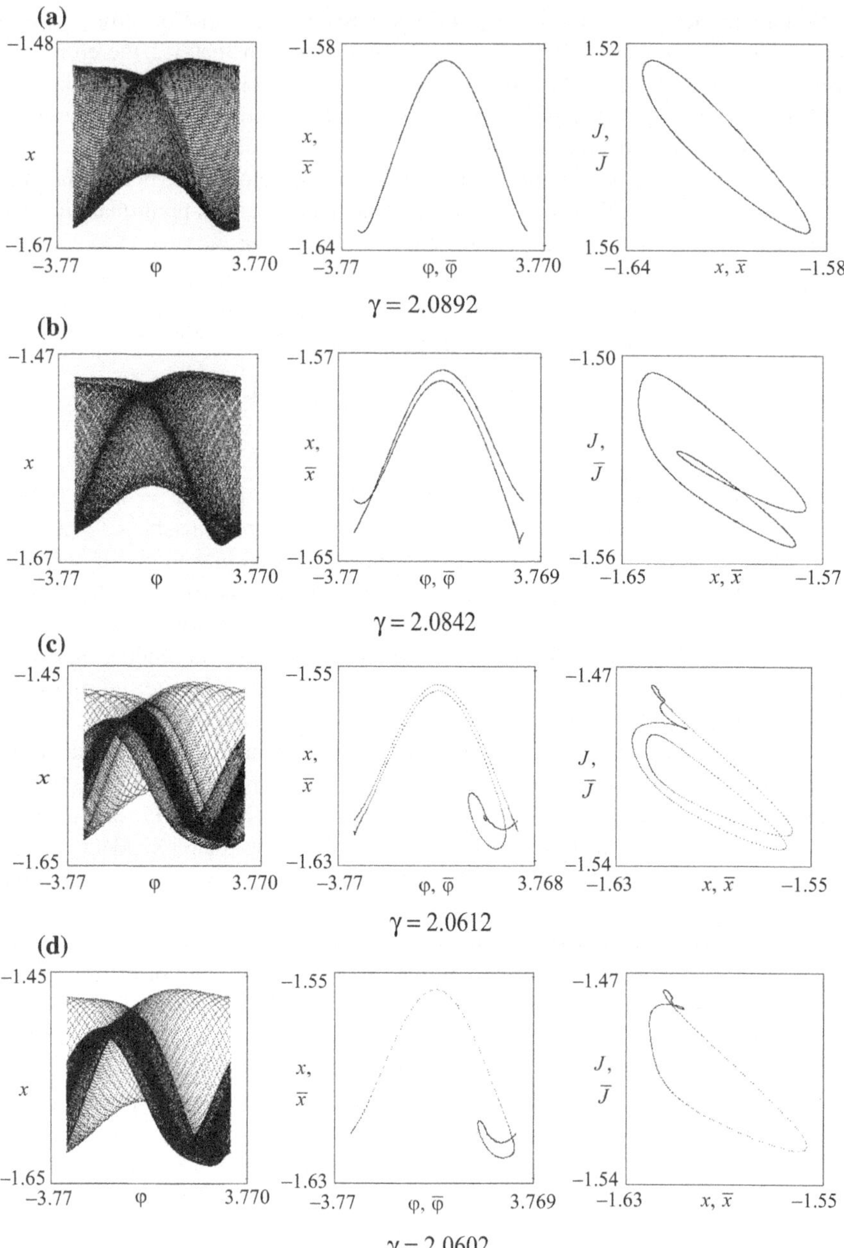

Fig. 2.8 The gallery phase portraits and portraits of point mappings obtained in numerical experiment 1 (continues on the next pages)

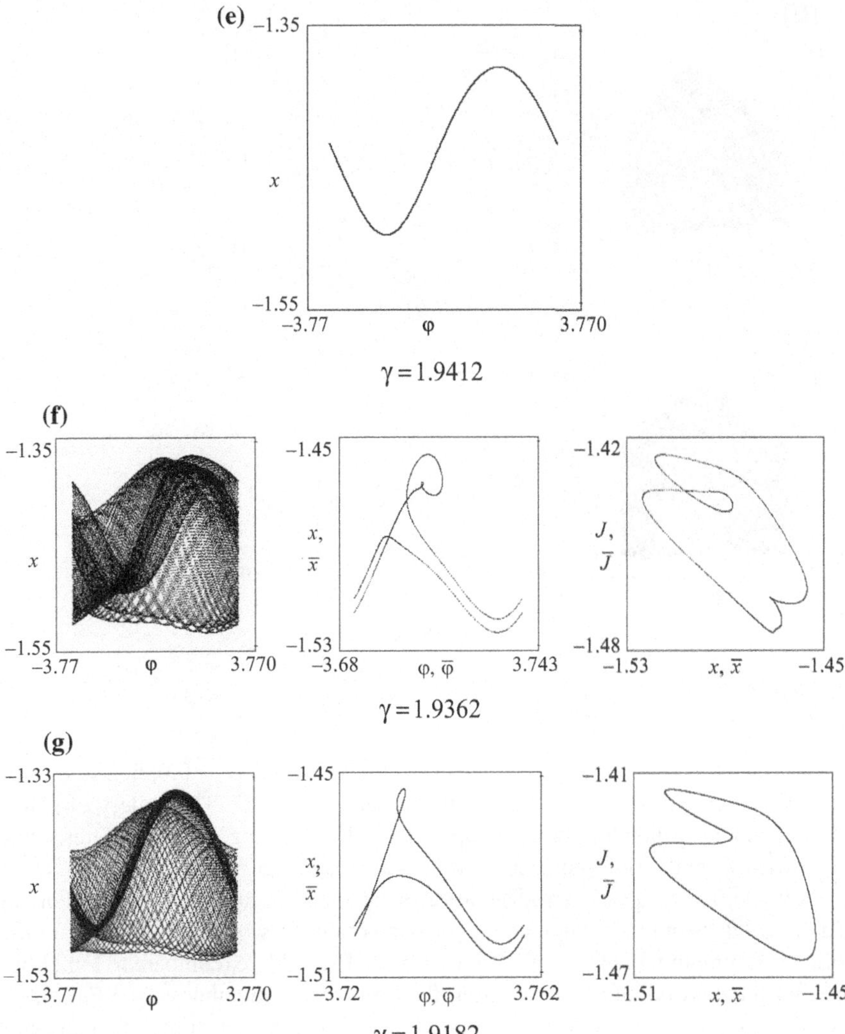

Fig. 2.8 (continued)

(h)

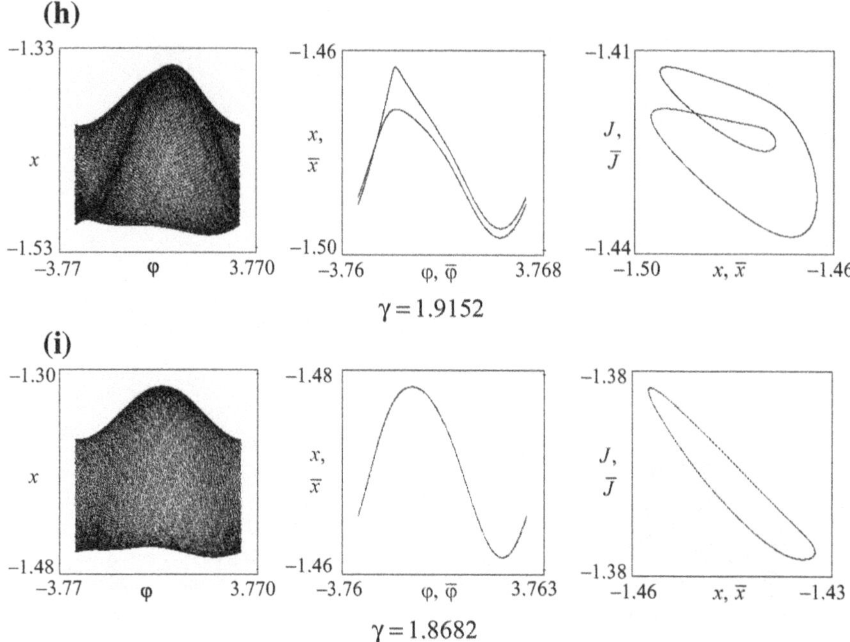

$\gamma = 1.9152$

(i)

$\gamma = 1.8682$

Fig. 2.8 (continued)

Experiment 2. The following parameter set has been used: $\mu = 0.086$, $h = 0.006$, $a = -0.006$, $b = 0.003$, $A = 2.5$, In this case, $\lambda^r = 0.517$. The gallery of phase portraits and corresponding point mappings are shown in Fig. 2.9. Parameter γ has been varied from the right-hand side towards the resonance step.

For $\gamma = 3.6091$, a quasi-periodic motion of the rotator occurs, see Fig. 2.9a. For the decreased values of this parameter, the torus experiences two pitchfork bifurcations, while itself remaining ergodic and "smooth", see Fig. 2.9b, c (compare to Fig. 2.6b, c). After the next pitchfork bifurcation, the torus loses smoothness, see Fig. 2.9d. A chain of pitchfork bifurcations leads to the origination of a strange Feigenbaum attractor, see Fig. 2.9e (compare to Fig. 2.6i). Further, a Shilnikov attractor merges with this attractor. For the decreased γ, a stable torus of the period four is born at the united attractor, see Fig. 2.9f. Further, the torus experiences a series of the opposite pitchfork bifurcations while itself remaining "smooth", see Fig. 2.9g–i. Finally, a torus of the period one loses stability and the rotator suddenly jumps into the synchronization regime, see Fig. 2.9j (this figure shows only a phase trajectory).

Experiment 3. In the previous experiment, during a series of pitchfork bifurcations the torus lost its smoothness. The question is: is it a rule or not? The following parameter set has been used: $\mu = 0.006$, $h = 0.004$, $a = -0.006$, $b = 0$, $A = 5$, $\omega_0 = 0.5$, In this case, $\lambda^r = 0.5$.

The gallery of phase portraits and corresponding point mappings are shown in Fig. 2.10. Parameter γ has varied towards the resonance step from the left-hand side.

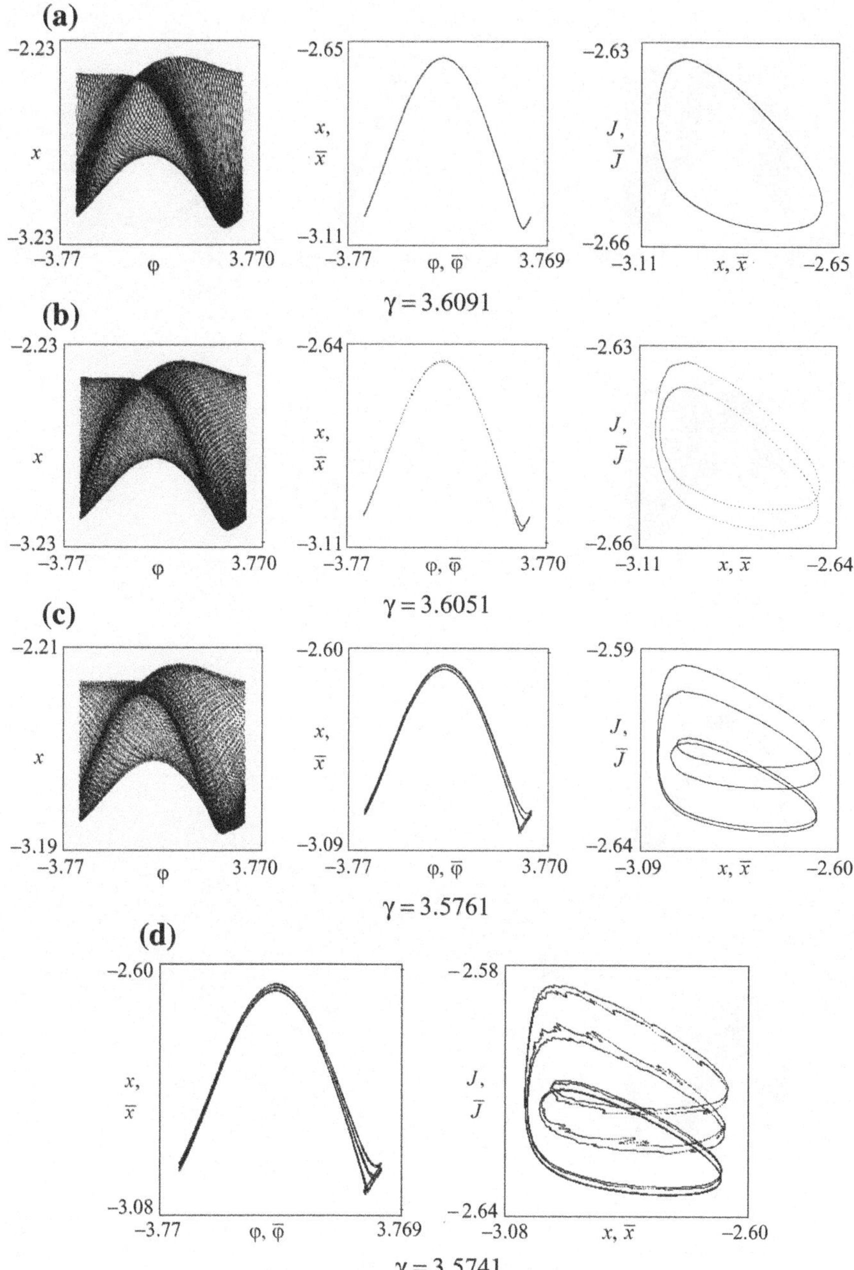

Fig. 2.9 The gallery phase portraits and portraits of point mappings obtained in numerical experiment 2 (continues on the next pages)

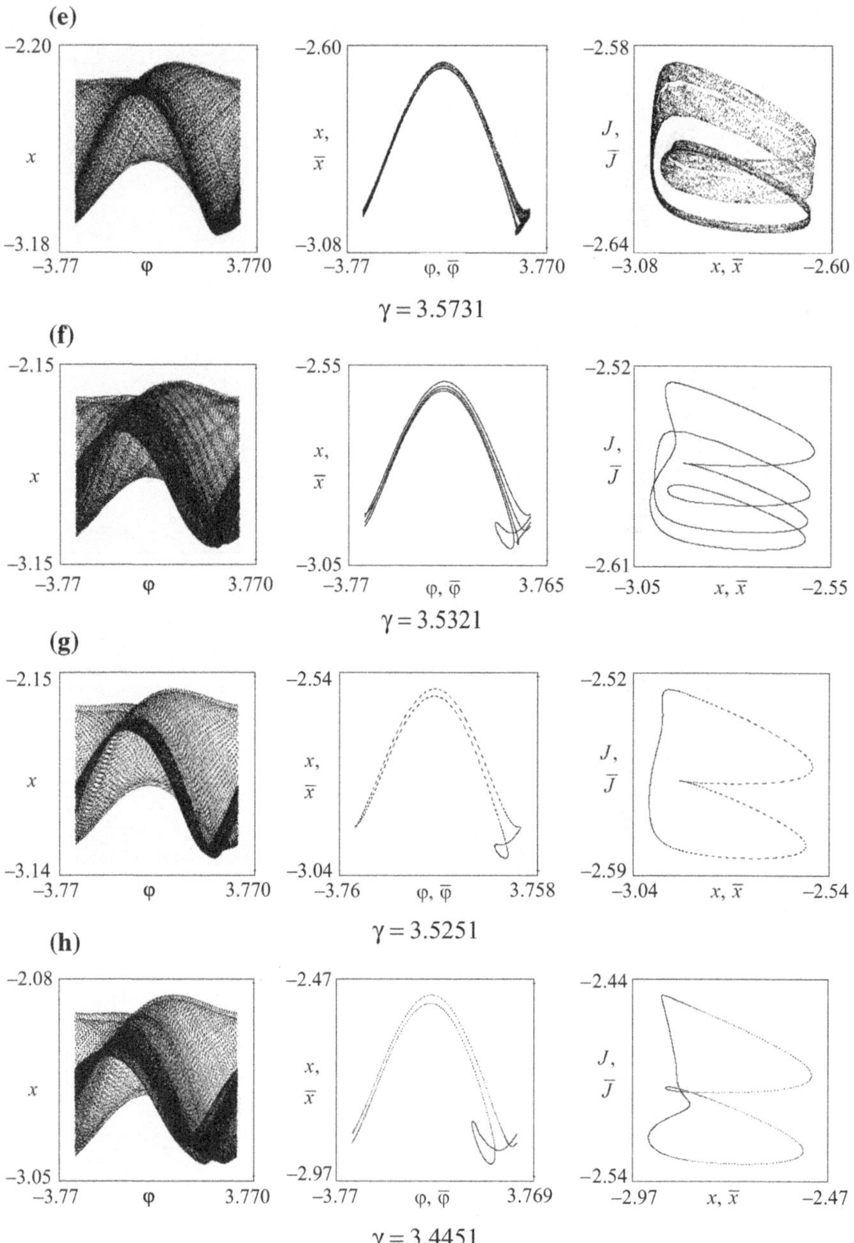

Fig. 2.9 (continued)

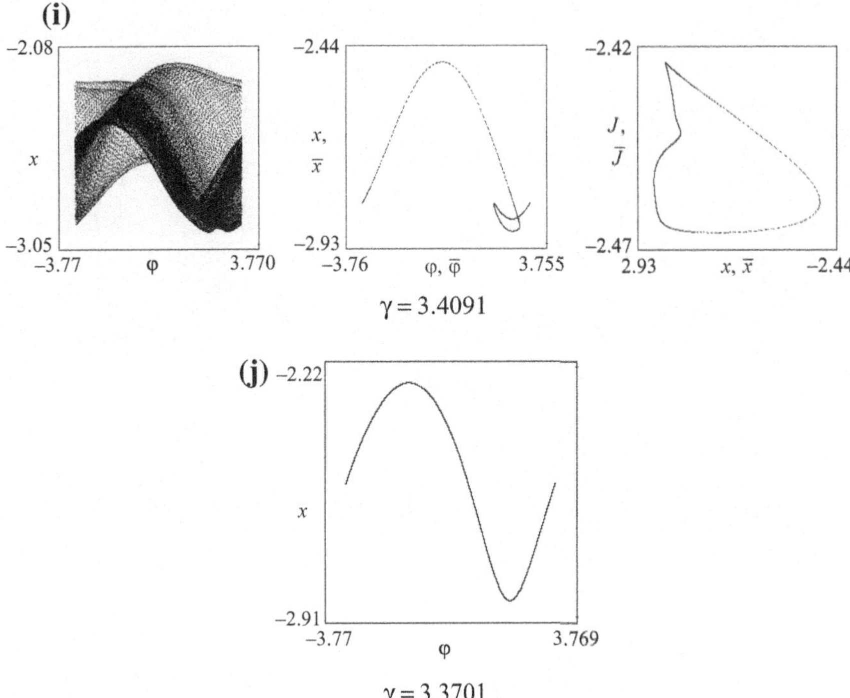

$$\gamma = 3.4091$$

$$\gamma = 3.3701$$

Fig. 2.9 (continued)

For $\gamma = 1.19534$ (see Fig. 2.10a), there exists a doubled smooth torus. During the increase of γ, the torus experiences a cascade of pitchfork bifurcations and does not lose its smoothness, see Fig. 2.10b, c. As a result of doublings of this smooth torus, a Feigenbaum attractor is originated, see Fig. 2.10d (compare to Fig. 2.6i).

Finalizing this section, we would like to draw the reader's attention to how the rotator returns to the "zero" resonance step, i.e. comes from the regime of rotations to the oscillatory regime. Figure 2.11a shows a trajectory of a complex rotary-oscillatory motion of the rotator and corresponding discrete trajectory of the point mapping (parameter γ has been varied while all other parameters have the same values as above). This trajectory represents a limit cycle. The dark part of it corresponds to a more dense concentration of trajectories. During the decrease of parameter γ, due to the trajectories thickening, a divisible oscillatory limit cycle originates. In the moment of bifurcation, a period of the limit cycle becomes infinite. Such a bifurcation is called the "blue sky catastrophe" [7, 8]. For further decrease of γ, the aforementioned limit cycle splits into two limit cycles: stable and unstable ones and the rotator moves along the stable limit cycle, see 2.11b.

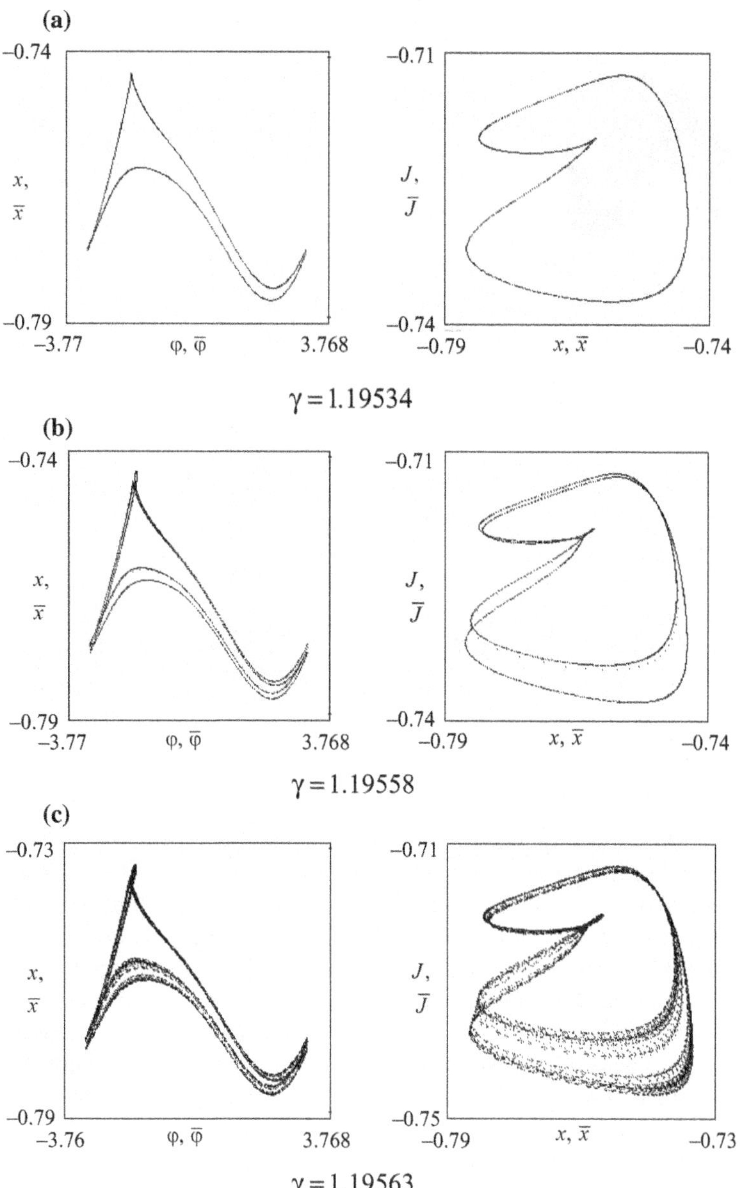

Fig. 2.10 The gallery phase portraits and portraits of point mappings obtained in numerical experiment 2 (continues on the next pages)

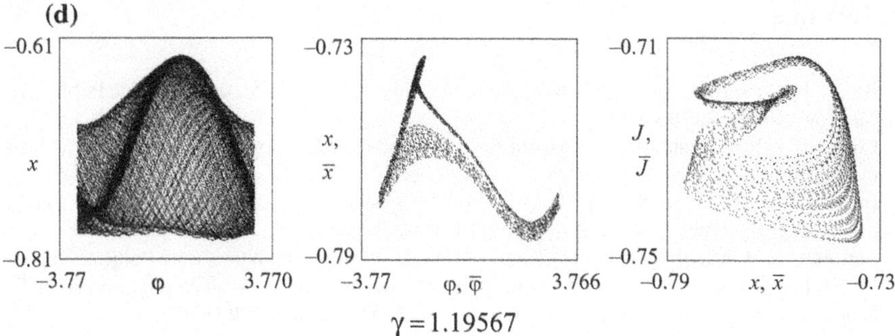

Fig. 2.10 (continued)

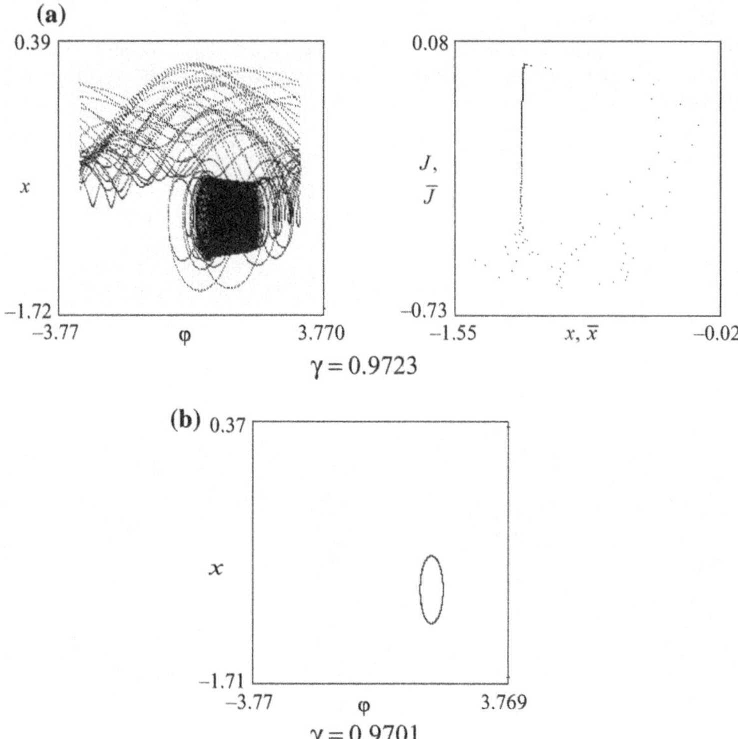

Fig. 2.11 Blue sky catastrophe bifurcation and oscillatory regime of rotator for $\mu = 0.086$, $h = 0.006$, $a = -0.006$, $b = 0.003$, $A = 2.5$, $\omega_0 = 0.8$

References

1. Birger, I.A., Panovko Ya, G. (eds.): Strength, Stability, Vibrations. Mashinostroenie Publishers, Moscow (1968). (in Russian)
2. Likharev, K.K.: Dynamics of Josephson Junctions and Circuits. Gordon and Breach, New York (1986)
3. Belykh, V.N., Nekorkin, V.I.: Qualitative study of a system of three differential equations in the theory of phase synchronization. Appl. Mathem. Mech. **39**(4), 642 (1975)
4. Shilnikov, L.P.: Theory of bifurcations of dynamical systems with homoclinic Poincare curves. In: VII International Conference for Nonlinear Oscillations. Bd. 12. Akad-Verlag, Berlin (1977)
5. Hale, J.K.: Oscillations in Nonlinear Systems. NewYork, McGrawHill (1963)
6. Mitropolsky, YuA, Lykova, O.V.: Integral Manifolds in Nonlinear Mechanics. Nauka, Moscow (1973). (in Russian)
7. Arnold, V.I. (ed.), Gamkrelidze, R.V. (ed.-in-chief), Progress in Science and Technology. Current Problems in Mathematics. Fundamental Directions 3 (Dynamical Systems III), VINITI AN SSSR, Moscow (1985)
8. Shilnikov, L.P., Turaev, D.V.: A new simple bifurcation of a periodic orbit of blue sky catastrophe type. Amer. Math. Soc. Transl. Ser. «Methods Qual. Theory Diff. Equat. Related Topics» **2**(200), 165–188 (2000)

Chapter 3
Autonomous Systems with Two Degrees of Freedom

3.1 Dynamics of Rotator with Oscillatory Load

Consider a model shown in Fig. 3.1. An asynchronous motor is coupled to the elastic shaft with an unbalanced disk (rotator) and balancing sleeve (oscillator) elastically coupled to the disk that can oscillate along the shaft. This sleeve represents the oscillatory load. We assume that the shaft is placed horizontally and the sleeve has no displacement along the shaft. There is viscous friction between the shaft and the sleeve.

Let us obtain the dynamic model of this system using the Lagrange function:

$$L = T + V = \frac{I_1}{2}\dot{\varphi}^2 + \frac{I_0}{2}\dot{\vartheta}^2 + mge(1 - \cos\varphi) + \frac{k}{2}(\varphi - \vartheta)^2,$$

where φ, ϑ are the angles of rotation of the shaft and the sleeve, respectively; I_1, I_0 are the moments of inertia of the disk and the sleeve, respectively; and k is the torsional stiffness of the spring.

Suppose that the torque is a linear function of the speed of rotation $M_d = M_0 - \lambda\dot{\varphi}$, and the moment of the friction force acting on the sleeve from the shaft has the form $\delta_0(\dot{\varphi} - \dot{\theta})$. In this case, the governing equations have the form

$$I_1\ddot{\varphi} + mge\sin\varphi + k(\varphi - \vartheta) = M_0 - \lambda\dot{\varphi},$$
$$I_0\ddot{\vartheta} + k(\vartheta - \varphi) = \delta_0(\dot{\varphi} - \dot{\vartheta}).$$

After the introduction of a new variable $\frac{k}{mge}(\vartheta - \varphi) = J$ and time $\frac{mge}{\lambda}\tau = \tau_n$, this system takes the form

$$I\ddot{\varphi} + \dot{\varphi} + \sin\varphi = \gamma + J,$$
$$\ddot{J} + \delta\dot{J} + \omega_0^2 J = b\ddot{\varphi}, \tag{3.1}$$

© Springer Nature Switzerland AG 2020

N. Verichev et al., *Chaos, Synchronization and Structures in Dynamics of Systems with Cylindrical Phase Space*, Understanding Complex Systems, https://doi.org/10.1007/978-3-030-36103-7_3

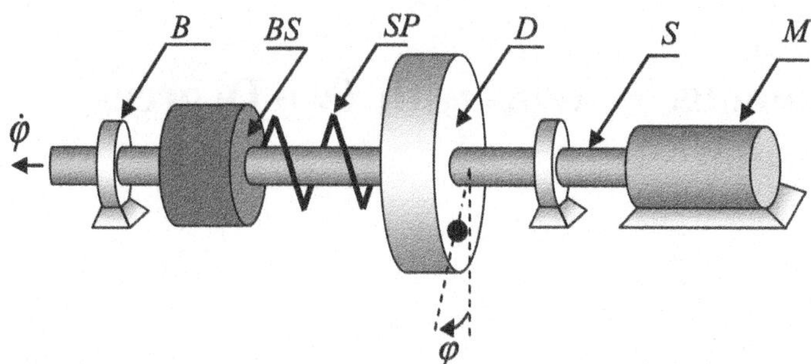

Fig. 3.1 Asynchronous motor with unbalanced disk elastically coupled to the balancing sleeve. M is the motor, S is the shaft, D is the disk, SP is the spring, BS is the balancing sleeve, and B is the bearing

where $I = \frac{I_1 mge}{\lambda^2}$, $\gamma = \frac{M_0}{mge}$, $\delta = \delta_0 I_0^{-1} \frac{\lambda}{mge}$, $\omega_0^2 = k I_0^{-1} \left(\frac{\lambda}{mge}\right)^2$, $b = -\frac{k}{mge}$. Note that variable J has a meaning of the angle of rotation of the sleeve against the shaft.

System (3.1) represents a system of the rotator-oscillator type [1]. Let us study this system for the parameter domain $D_\mu = \left\{I^{-1} = \mu \ll 1, \delta = \mu h, \gamma - \omega_0 = \mu \Delta\right\}$, which physically corresponds to a quasi-linear rotator that interacts resonantly with a high quality oscillator. Let us reduce this system to the standard form having considered the equivalent system of the form

$$I \ddot{\varphi} + \dot{\varphi} + \sin \varphi = \gamma + J,$$
$$\dot{J} = W + b \dot{\varphi} - \mu h J,$$
$$\dot{W} = -\omega_0^2 J. \tag{3.2}$$

Let us make a modified van der Pol change of the variables in the equations for the oscillator, and apply already known transformations for the equation for the rotator:

$$J = \theta \sin \varphi + \eta \cos \varphi,$$
$$W = (\theta \cos \varphi - \eta \sin \varphi)\omega_0 - b\omega_0,$$
$$\dot{\varphi} = \omega_0 + \mu \Phi(\theta, \eta, \varphi, \xi). \tag{3.3}$$

This change of the variables transforms equations of the oscillator to the equivalent system of the form

$$\dot{\theta} = \mu \Theta(\theta, \eta, \xi, \varphi),$$
$$\dot{\eta} = \mu T(\theta, \eta, \xi, \varphi),$$

where $\Theta = (\eta + b \sin \varphi)\Phi - h(\theta \sin \varphi + \eta \cos \varphi) \sin \varphi$, $T = (-\theta + b \cos \varphi)\Phi - h J \cos \varphi$.

Substitution of the last equation of system (3.3) into the first equation of system (3.2) results in the following equation:

$$\frac{\partial \Phi}{\partial \theta}\mu\Theta + \frac{\partial \Phi}{\partial \eta}\mu T + \frac{\partial \Phi}{\partial \varphi}\omega_0 + \frac{\partial \Phi}{\partial \varphi}\mu\Phi + \mu\Phi + \dot{\xi} = \mu\Delta + (\theta - 1)\sin\varphi + \eta\cos\varphi,$$

from which we obtain equations that define functions $\Phi(\varphi, \theta, \eta, \xi)$, $\Xi(\varphi, \theta, \eta, \xi)$, from which we obtain these functions:

$$\frac{\partial \Phi}{\partial \varphi}\omega_0 = (\theta - 1)\sin\varphi + \eta\cos\varphi, \quad \Phi = \frac{1-\theta}{\omega_0}\cos\varphi + \frac{\eta}{\omega_0}\sin\varphi + \xi,$$

$$\Xi = \Delta - \left(\frac{\partial \Phi}{\partial \theta}\Theta + \frac{\partial \Phi}{\partial \eta}T + \frac{\partial \Phi}{\partial \varphi}\Phi + \Phi\right).$$

The final result of transformations is a system in the standard form, which is equivalent to system (3.1):

$$\dot{\theta} = \mu\Theta(\theta, \eta, \xi, \varphi),$$
$$\dot{\eta} = \mu T(\theta, \eta, \xi, \varphi),$$
$$\dot{\xi} = \mu\Xi(\theta, \eta, \xi, \varphi),$$
$$\dot{\varphi} = \omega_0 + \mu\Phi(\theta, \eta, \xi, \varphi). \tag{3.4}$$

Having averaged (3.4) over a fast spinning phase, we obtain a truncated system of the form

$$\dot{\xi} = \mu(-\xi + b_3\eta + \Delta),$$
$$\dot{\theta} = \mu(-b_4\theta + b_5\eta + \eta\xi),$$
$$\dot{\eta} = \mu(-b_4\eta - b_5\theta - \theta\xi + b_7),$$
$$\dot{\varphi} = \omega_0 + \mu\xi, \tag{3.5}$$

where $b_3 = \frac{1}{2\omega_0^2}$, $b_4 = \frac{h}{2}$, $b_5 = \frac{b}{2\omega_0}$, $b_7 = \frac{b}{2\omega_0}$.

Then, by applying changes of the variables and time of the form

$$x = b_4^{-1}(\xi + b_5), \quad y = \frac{b_3}{b_4}\eta - \Lambda, \quad z = \frac{b_3}{b_4}\theta, \quad \mu b_4\tau = \tau_n$$

we reduce the first three equations of system (3.5) to a system of the form

$$\dot{x} = -\sigma(x - y) + \rho,$$
$$\dot{y} = -y - xz,$$
$$\dot{z} = -z + xy + \Lambda x, \tag{3.6}$$

where $\sigma = \frac{1}{b_4}$, $\rho = \frac{(\Delta + (b_4 b_5 + b_3 b_7)/b_4)}{b_4^2}$, $\Lambda = \frac{b_3 b_7}{b_4^2}$.

Let us discuss some of the dynamical properties of the averaged system (3.6).

1. The system is a dissipative system. This property is justified with the help of quadratic form $V = \frac{1}{2}(x^2 + (y + \Lambda)^2 + (z - \sigma)^2)$, whose derivative taken along the trajectories of system (3.6) has the form $\dot{V} = -\sigma x^2 - (\rho - \sigma \Lambda)x - y^2 - \Lambda y - z^2 + \sigma z$. It is obvious that it is negative outside a sphere $V \leq L^2$. This means that equilibria, limit cycles, and other limit trajectories (if such ones exist) remain inside a bounded spherical domain of the phase space.

2. Depending on the parameters, this system has up to three equilibria with coordinates

$$x_0 = \omega, \quad y_0 = -\frac{\Lambda \omega^2}{1 + \omega^2}, \quad z_0 = \frac{\Lambda \omega}{1 + \omega^2},$$

where $\omega_{1,2,3}$ are the solutions of equation

$$f = \omega + \frac{\Lambda \omega^2}{1 + \omega^2} = \frac{\rho}{\sigma}. \tag{3.7}$$

Note that the aforementioned parameter ω is proportional to the dynamical mistuning of frequencies of the rotator and oscillator. We will call curve $f = f(\omega)$ the curve of equilibria.

3. If system (3.6) has one equilibrium $O(x_0, y_0, z_0)$, then this equilibrium is globally asymptotically stable. This can be proven with the help of the Lyapunov function $V = \frac{1}{2}(mx_1^2 + y_1^2 + z_1^2)$, where $x_1 = x - x_0$, $y_1 = y - y_0$, $z_1 = z - z_0$, the derivative of which, taken along the vector field of the system, has the form $\dot{V} = -(\alpha_1 x_1 + y_1)^2 - (\alpha_2 x_1 + z_1)^2 \leq 0, \forall(x_1, y_1, z_1)$, where $\alpha_1 = \frac{\sigma m - z_0}{2}$, $\alpha_2 = \frac{-(y_0 + \Lambda)}{2}$, and m is a positive root of equation $\sigma^2 m^2 - 2\sigma(z_0 + 2)m + z_0^2 + (y_0 + \Lambda)^2 = 0$. It may be proven that conditions of existence of a positive root and conditions of unicity of the equilibrium in system (3.6) (conditions that function f is a one-one function) are the same. It is not difficult to find that f is a one-one function for $|\Lambda| \leq \frac{8\sqrt{3}}{9}$.

4. Stability of equilibria. A characteristic equation for an arbitrary equilibrium $O(x_0, y_0, z_0)$ of system (3.6) has the form

$$p^3 + a_0 p^2 + a_1 p + a_2 = 0,$$

where $a_0 = \sigma + 2$, $a_1 = 2\sigma + 1 + \omega^2 + \Lambda \sigma \frac{\omega}{1 + \omega^2}$, $a_2 = \sigma(1 + \omega^2) \times \left(1 + 2\Lambda \frac{\omega}{(1 + \omega^2)^2}\right)$. If $\omega = 0 \, (\rho = 0)$, then equilibrium $O(0, 0, 0)$ is stable. If $\omega \neq 0$, then the Gurvitz conditions are equivalent to a set of the following inequalities:

$$a_1 > 0 : f > (<) f_1,$$

$$a_2 > 0 : f > (<) f_2,$$

$$a_0 a_1 - a_2 > 0 : f > (<) f_3 \quad \text{for } \omega > 0 \ (\omega < 0),$$

where $f_1 = -\omega \frac{\sigma + 1 + \omega^2}{\sigma}$, $f_2 = \frac{\omega(1 - \omega^2)}{2}$, $f_3 = -\omega \frac{\sigma^2 + 4\sigma + 2 + 2\omega^2}{\sigma^2}$. Thus, the criterion of stability of any equilibrium of the system consists of a location of a point of the curve, corresponding to equilibria (3.7), above (below) all curves $f_{1,2,3}$ for $\omega > 0$ $(\omega < 0)$. In the parameter space (σ, Λ, ρ) of equation $f = f_{1,2,3}$, $f = \rho/\sigma$ define bifurcation surfaces, at which changes of the character of stability of equilibriums take place.

The auxiliary functions have two simple properties:

(1) Function f_2 crosses f in the positions of extrema and at the origin;

(2) All curves cross each other at three points, $\omega_{1,2} = \pm\sqrt{\frac{3\sigma + 2}{\sigma - 2}}$. and at the origin.

The aforementioned information regarding the properties of system (3.6) will be discussed in more detail during the consideration of the rotation characteristic of the rotator.

Rotation characteristic of rotator. As usual, our interest is concerned with qualitative forms of the rotation characteristic, i.e. with the study of different types of rotator's limit motions. Namely, we are interested in the dependency of the rotational speed on the constant component of torque (γ), while all other parameters are considered to be constant.

According to definition of the rotation characteristic and due to the principle of the averaging as well as due to the form of the last equation of system (3.5), we obtain $\langle \dot{\varphi} \rangle_\tau = \omega_0 + \mu \langle \xi(\tau) \rangle_\tau$. This equation determines the rotation characteristic in the resonance zone of the oscillatory system, i.e. in the vicinity of frequency ω_0, since $\gamma \sim \Delta \sim \rho$, and $\xi(t) \sim x(t)$ (due to changes of parameters and variables). Thus, we have obtained that the rotation characteristic and curve $\rho = \rho(x^*(t, t_0))$ have similar qualitative features. One curve can be simply scaled into another one. The latter curve will be considered as a rotation characteristic. Here, $x^*(t, t_0)$ are the solutions of averaged system (3.6) that correspond to limit sets of its trajectories: equilibria, limit cycles, tori, and strange attractors.

Consider in more detail the curve of equilibria that represents a part of the rotation characteristic. It is a one-one function, which is stable for $|\Lambda| \leq \frac{8\sqrt{3}}{9}$ (see Fig. 3.2a). For $|\Lambda| > \frac{8\sqrt{3}}{9}$, this curve goes through the origin and has two extrema in the right-hand half-plane (see Fig. 3.2b, c). It is easy to see that if $\frac{df}{d\omega} < 0$, then $a_2 < 0$, thus all points of the part of the curve that has a negative slope (between the extrema) are unstable and correspond to equilibria of saddle or saddle-focus types.

Stability or instability of other parts of the curve of equilibria is established through the analysis of behavior of auxiliary functions. One can see that curve f_2 goes through the extrema and determines instability of the negatively sloped part. For the points of positively sloped parts, inequalities (conditions of stability) for functions f and

Fig. 3.2 Qualitative forms of the rotation characteristic of rotator: **a** $\sigma = 5$, $\Lambda = -1.2$; **b** $\sigma = 5$, $\Lambda = -7$; **c** $\sigma = 5$, $\Lambda = -25$

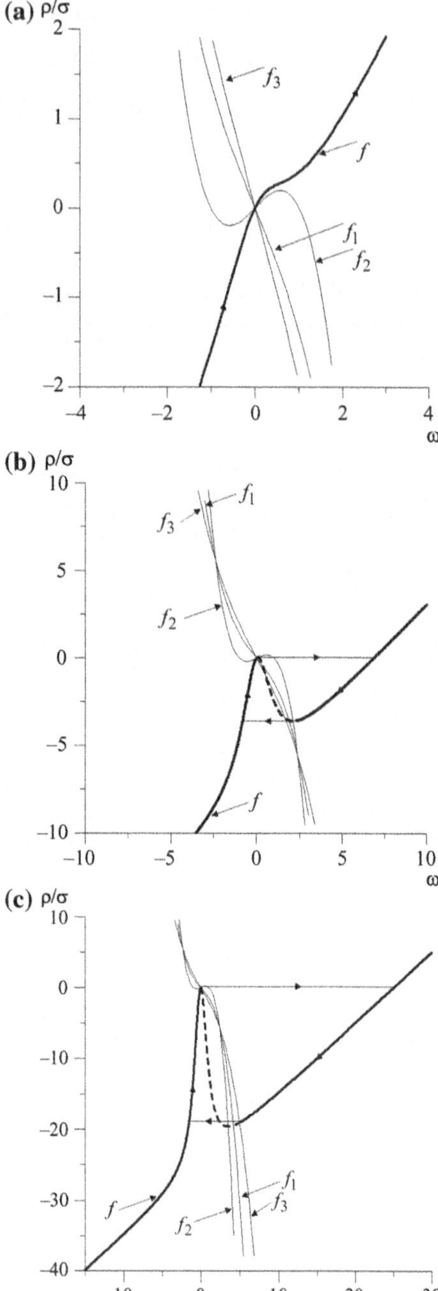

f_2 are fulfilled. Therefore, everything depends on a mutual distribution of functions f and f_1, as well as f and f_3. For the points of the positively sloped part in the left-hand half-plane ($\omega < 0$) $f_1 > 0$, $f_3 > 0$, but $f < 0$, so that stability conditions are fulfilled. On the contrary, for the points of the positively sloped part in the right-hand half-plane, the opposite takes place: $f_1 < 0$, $f_3 < 0$, but $f > 0$ and corresponding inequalities are also fulfilled. Thus, all points of the left-hand positively sloped part of curve $f = f(\omega)$ correspond to stable equilibria.

There exist two special cases for the right positively sloped part of the curve:

(1) All of its points are stable (see Fig. 3.2b);
(2) a section of the positively sloped part of the curve, which is adjacent to the minimum, corresponds to unstable equilibria (see Fig. 3.2c). Parameters for both cases are obtained through the estimation of location of the point of intersection of curves f_1, f_2, f_3 with respect to curve $f = f(\omega)$. If this point is located below the curve, then conditions of stability are fulfilled for all of the points of the positively sloped part. If this point is placed above the curve, then the positively sloped part will be unstable. Thus, inequality $f(\omega_1) < f_2(\omega_1)$ defines the second case, while the opposite inequality defines the first case. The explicit form of this inequality for actual parameters of the system under consideration is as follows: $\Lambda < -\frac{8\sigma^2}{(\sigma-2)^2}\sqrt{\frac{\sigma-2}{3\sigma+2}}$.

Bifurcations of equilibria. There exist two types of bifurcations in the system:

(1) Bifurcation of merging of the equilibria, i.e. formation of a saddle-node and its disappearance in the points of the extremum of function $f = f(\omega)$;
(2) Inversed Andronov-Hopf bifurcation for the equilibria that correspond to the right-hand positively sloped part of curve $f = f(\omega)$, when an unstable limit cycle, which originates from the separatrix loop of a saddle equilibrium, gets stuck in the stable equilibrium and "transfers" its own instability to this equilibrium.

There have been found no limit sets of trajectories other than equilibria in system (3.6). The dynamics of the system under consideration is quite simple. Suppose that a constant component of the torque is increased quasi-statically, see Fig. 3.2b. In this case, the frequency of rotation of the rotor (i.e. of the motor) is increased smoothly and it slows down as soon as it approaches the natural frequency of the oscillatory system, i.e. energy of the motor is transferred to vibrations rather than to increase the frequency of rotation. When the torque reaches the value that corresponds to the maximum of curve $f = f(\omega)$, the speed of rotation suddenly jumps up to the value that corresponds to the right-hand positively sloped branch. For the opposite variation of rpm (the process of slowing the motor down), such jump occurs from the minimum of curve $f = f(\omega)$. This means that the classical Sommerfeld effect occurs in the system. The only difference from the case shown in Fig. 3.2 consists of the fact that during the shutdown process, a frequency jump occurs not from the minimum of the curve $f = f(\omega)$, but from a point of the positively sloped part of this curve, so that a truncated hysteretic loop is observed in this case.

3.2 Chaotic Dynamics of a Simple Vibrational Mechanism

Consider the dynamics of a simple vibrational mechanism shown in Fig. 3.3. A mass m is placed over a conveyor belt. It is attached a the stiff wall by a visco-elastic coupling with stiffness c and viscosity k. The crankshaft of length r_1 is placed perpendicular to the shaft of the asynchronous motor and attached to the mass by the spring with stiffness c_1. The crankshaft length is small enough that the deformation of the elastic coupling can be considered in the horizontal direction only. The friction between the mass and the conveyor belt is viscous with coefficient q. The driving roller has an imbalance with mass m_0, placed at angle m_0, with respect to the crankshaft in the direction of the shaft's rotation.

Various modifications of this system are known from literature [2–5]. In what follows, we will perform an analytical study of this system in the framework of methods described in the previous sections to demonstrate that even such a simple system can perform chaotic vibrations. We will also discuss mechanisms of origination of dynamical chaos [6].

The governing equations of the system under consideration have the form

$$\ddot{x} + \omega_0^2 x = \frac{c_1 r_1}{m} \sin \varphi + \frac{qr}{m} \dot{\varphi} - \frac{q+k}{m} \dot{x},$$
$$I\ddot{\varphi} = \widetilde{M}_d(\dot{\varphi}) - rq(r\dot{\varphi} - \dot{x}) + c_1 r_1 (x - r_1 \sin \varphi) \cos \varphi - M_0 \cos(\varphi + \varphi_0). \tag{3.8}$$

Here $\omega_0^2 = (c + c_1)/m$, I is the normalized moment of inertia of the rotor, $\widetilde{M}_d(\dot{\varphi})$ is the motor torque including the moment of resistance forces acting on the rotor, $M_0 = m_0 g e$, and e is the eccentricity.

We will assume that the variables, parameters, and time in Eq. (3.8) are dimensionless. Consider the following parameter condition: $I^{-1} = \mu \ll 1$, where μ is a small parameter, $(q + k)/m = 2\mu h$ (dissipation in the "oscillatory" part of the system is sufficiently small), $c_1 r_1/m = 2\mu\lambda\omega_0$, $c_1 r_1 = 2\mu b\omega_0$. The condition of smallness, however, is not applied to other combinations of parameters.

As usual, we assume that $\widetilde{M}_d(\dot{\varphi}) = M_d - \delta\dot{\varphi}$, where M_d is the constant component of the torque, and δ is the generalized coefficient of resistance forces.

For the aforementioned parameters, dynamical system (3.8) exemplifies a rotator with an oscillatory load. Similarly to Sect. 3.1, this system is studied in a zone of

Fig. 3.3 A scheme of a simple vibrational mechanism

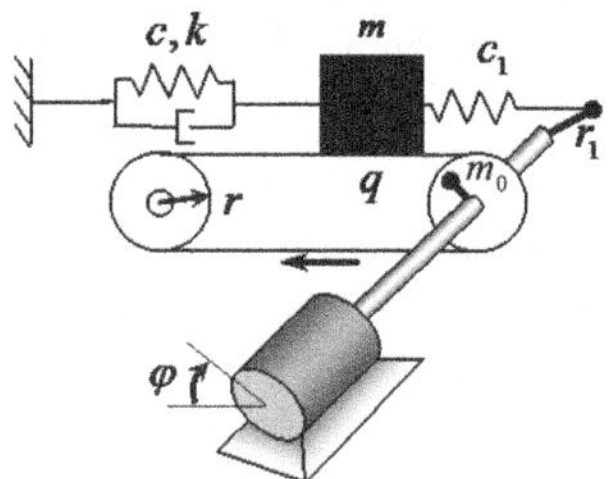

the main resonance. In this case, the resonance zone is defined by equation $M_d - (\delta + r^2 q)\omega_0 = \mu\Delta$.

Let us reduce system (3.8) to an equivalent system in a standard form

$$\dot{\theta} = \mu F_1(\theta, \eta, \varphi, \xi),$$
$$\dot{\eta} = \mu F_2(\theta, \eta, \varphi, \xi),$$
$$\dot{\xi} = \mu F_3(\theta, \eta, \varphi, \xi),$$
$$\dot{\varphi} = \omega_0 + \mu \Phi(\theta, \eta, \varphi, \xi) \qquad (3.9)$$

By applying the following changes of the variables:

$$x = qr/m\omega_0 + \theta \sin\varphi + \eta \cos\varphi,$$
$$\dot{x} = (\theta \cos\varphi - \eta \sin\varphi)\omega_0,$$
$$\Phi = rq(\theta \sin\varphi + \eta \cos\varphi) - M_0/\omega_0 \sin(\varphi + \varphi_0) + \xi.$$

In system (3.9)

$$F_1 = \eta\Phi + 1/\omega_0 X(.)\cos\varphi,$$
$$F_2 = -\theta\Phi - 1/\omega_0 X(.)\sin\varphi,$$

$$X(.) = 2\lambda\omega_0 \sin\varphi + qr/m\Phi - 2h\omega_0(\theta \cos\varphi - \eta \sin\varphi),$$

$$F_3 = -\left(\frac{\partial\Phi}{\partial\theta}F_1 + \frac{\partial\Phi}{\partial\eta}F_2 + \frac{\partial\Phi}{\partial\varphi}\Phi\right) - (\delta + qr^2)\Phi$$
$$+ 2b\omega_0(\theta \sin\varphi + \eta \cos\varphi - r_1 \sin\varphi)\cos\varphi.$$

Having averaged system (3.9) over a fast-spinning phase [7], we obtain a truncated system of the form

$$\dot{\xi} = \mu(-b_1\xi + b_2\theta + b_3\eta + \Delta),$$
$$\dot{\theta} = \mu(-b_4\theta + b_5\eta + \eta\xi + b_6),$$
$$\dot{\eta} = \mu(-b_4\eta - b_5\theta - \theta\xi + b_7),$$
$$\dot{\varphi} = \omega_0 + \mu\xi, \qquad (3.10)$$

where $b_1 = qr^2 + \delta$, $b_2 = -\dfrac{M_0 rq}{2\omega_0 \sin\varphi_0}$, $b_3 = b\omega_0 + \dfrac{M_0 rq}{2\omega_0 \cos\varphi_0}$, $b_4 = h$, $b_5 = \dfrac{(qr)^2}{2m\omega_0}$, $b_6 = -\dfrac{M_0 rq}{2m\omega_0^2 \sin\varphi_0}$, $b_7 = \dfrac{M_0 rq}{2m\omega_0^2 \cos\varphi_0 - \lambda}$.

Then, using a change of variables and time of the form

$$x = b_4^{-1}(\xi + b_5), \quad y = (b_3\eta + b_2\theta)/b_1 b_4 - \Lambda,$$
$$z = (b_3\theta - b_2\eta)/b_1 b_4 + R, \quad \mu b_4\tau = \tau_n$$

we reduce the first three equations of system (3.10) to the following system

$$\dot{x} = -\sigma(x - y) + \rho,$$
$$\dot{y} = -y + Rx - xz,$$
$$\dot{z} = -z + xy + \Lambda x, \tag{3.11}$$

where

$$\sigma = b_1/b_4 = (qr^2 + \delta)/h, \quad \rho = \frac{\Delta + (b_1 b_4 b_5 + b_3 b_7 + b_2 b_6)/b_4}{b_4^2},$$

$$R = \frac{b_2 b_7 - b_3 b_6}{b_1 b_4^2} = \frac{2 A \lambda}{(\delta + qr^2)h^2} \sin \varphi_0, \quad A = \frac{M_0 qr}{2\omega_0},$$

$$\Lambda = \frac{b_3 b_7 + b_2 b_6}{b_1 b_4^2} = \frac{\lambda(A^2 - b^2 \omega_0^2)}{b\omega_0(\delta + qr^2)h^2}.$$

As before, our interest is mainly concerned with the study of different types of limit sets in the averaged system and, corresponding to them, qualitative forms of rotation characteristics in the resonance zone. After the application of the new variables and parameters, it is found that $M_d \sim \Delta \sim \rho$, $\xi(t) \sim x(t)$ (these parameters have linear dependencies, similar to the case described in Sect. 3.1). Therefore, curve $\rho = \rho(x^*(t, t_0))$, has the same qualitative features as the rotation characteristic in a resonance zone.

Let us list properties of system (3.11). Note that some of these properties coincide with properties of the system studied in Sect. 3.1.

1. System (3.11) is dissipative. This property can be proven with the help of the following quadratic form: $V = \frac{1}{2}(x^2 + (y + \Lambda)^2 + (z - \sigma - R)^2)$. Its derivative, taken along trajectories of the system, has the form $\dot{V} = -\sigma x^2 - (\rho - \sigma\Lambda)x - y^2 - \Lambda y - z^2 + (\sigma + R)z$. It is clear that this quadratic form is negative for a sphere $V \leq L^2$, i.e. all limit sets of the trajectories of system (3.11) are limited by a sphere of dissipation.
2. Depending on the parameters, system (3.11) can have up to three equilibria with coordinates

$$x_0 = \omega, \quad y_0 = \frac{R\omega - \Lambda\omega^2}{1 + \omega^2}, \quad z_0 = \frac{R\omega^2 + \Lambda\omega}{1 + \omega^2},$$

where $\omega_{1,2,3}$ are the solutions of equation

$$f = \omega + \frac{\Lambda\omega^2 - R\omega}{1 + \omega^2}, \quad f = \rho/\sigma. \tag{3.12}$$

As before, parameter ω has a meaning of the mistuning between the frequency of the rotator's periodical rotations and the eigenfrequency of the oscillator.

3. If system (3.11) has one equilibrium $O(x_0, y_0, z_0)$, then it is globally asymptotically stable. This can be proven with the help of the Lyapunov function $V = \frac{1}{2}(mx_1^2 + y_1^2 + z_1^2)$, where $x_1 = x - x_0$, $y_1 = y - y_0$, $z_1 = z - z_0$. The derivative of the Lyapunov function, taken along trajectories of the system $\dot{V} = -(\alpha_1 x_1 + y_1)^2 - (\alpha_2 x_1 + z_1)^2 \leq 0$, $\forall (x_1, y_1, z_1)$, where $\alpha_1 = \frac{\sigma m + R - z_0}{2}$, $\alpha_2 = -\frac{y_0 + \Lambda}{2}$, and m is a positive root of the equation $\sigma^2 m^2 + 2\sigma(R - z_0 - 2)m + (R - z_0) + (y_0 + \Lambda)^2 = 0$. It may be proven that the conditions of existence of a positive root and conditions of unicity of the equilibrium in system (3.11) (conditions that function f is a one-one function) are exactly the same.

4. Properties of the equilibria. The characteristic equation of system (3.11) for an arbitrary equilibrium $O(x_0, y_0, z_0)$ has the form $p^3 + a_0 p^2 + a_1 p + a_2 = 0$, where $a_0 = \sigma + 2$, $a_1 = 2\sigma + 1 + \omega^2 + \sigma \frac{\Lambda\omega - R}{1 + \omega^2}$, $a_2 = \sigma(1 + \omega^2)\left(1 + \frac{R\omega^2 + 2\Lambda\omega - R}{(1 + \omega^2)^2}\right)$. For $\omega = 0\,(\rho = 0)$, the stability of equilibrium $O(0, 0, 0)$ is defined by inequality $R < 1$. For $\omega \neq 0$, the Gurvitz conditions are equivalent to the following inequalities:

$$a_1 > 0 : f > (<)f_1, \quad a_2 > 0 : f > (<)f_2,$$
$$a_0 a_1 - a_2 > 0 : f > (<)f_3, \quad \omega > 0(\omega < 0),$$

where $f_1 = -\omega(\sigma + 1 + \omega^2)/\sigma$, $f_2 = -\omega(R - 1 + \omega^2)/2$, $f_3 = -\omega(\sigma^2 + 4\sigma - \sigma R + 2 + 2\omega^2)/\sigma^2$. In other words, the criterion of the stability of any equilibrium of the system is a location of a point of curve (3.12) that corresponds to this equilibrium above (or below) all curves $f_{1,2,3}$ for $\omega > 0(\omega < 0)$. Equations $f = f_{1,2,3}$, $f = \rho/\sigma$ define bifurcation surfaces in the parameter space of the system, for the points of which the changes of structure of phase trajectories at local manifolds of the equilibria take place, as well and as changes in their stability.

As in Sect. 3.1, the auxiliary functions have two useful properties.

(a) f_2 crosses f at the origin and in the points of extremum;

(b) All curves cross each other at two points $\omega_{1,2} = \pm\sqrt{\frac{3\sigma - \sigma R + 2}{\sigma - 2}}$. and at the origin.

We will use these properties to study the stability of equilibria.

5. Chaotic attractors.

5.1. *Lorenz Attractor.* For $\rho = \Lambda = 0$, Eq. (3.11) represent a well-known Lorenz system [8]. This classical system, for the values of the parameters $\sigma = \sigma^*$, $R = R^* > R_c$, $R_c = \frac{\sigma^2 + 4\sigma}{\sigma - 2}$ has a unique attracting limit set in the phase space $G(x, y, z)$, a strange attractor (Lorenz attractor). Note that, according to property 4, the expression for R_c is defined by the condition of the coincidence of the zeros of functions f and f_3. This condition corresponds to the loss of stability of the equilibria $O_{2,3}(\pm\sqrt{R - 1}, \pm\sqrt{R - 1}, R - 1)$ caused by the fact that those saddle limit cycles get stuck in them. These limit cycles have been earlier born (by varying parameter R) from the separatrix loop of saddle $O_1(0, 0, 0)$. In fact, a strange attractor already exists for $R^* < R < R_c$, where R^* is the value of the parameter R that corresponds to the separatrix loop of the saddle. In this case, the strange attractor and stable equilibria coexist, having non-overlapping attraction domains. In this interval of values of parameter R, depending on initial conditions, either a strange attractor or one of two stable equilibria would be born [8].

"Deformation" and degeneration of the Lorenz attractor for nonzero parameters ρ and Λ have been studied in a numerical experiment. For $\Lambda = 0$, and by increasing parameter ρ from zero, the attractor loses its symmetry since the affix remains mostly in the vicinity of the right-hand saddle-focus (for the projection onto (x, z) plane. Further, this equilibrium becomes stable after the birth of a saddle limit cycle. In this case, depending on the initial conditions, either equilibrium or a chaotic attractor may originate. With further increase of parameter ρ, the saddle cycle gets "stuck" in the separatrix loop of a saddle and the equilibrium becomes globally stable. Due to invariance of system (3.11) with respect to a change of the variables of the form $(x, y, z) \mapsto (-x, -y, z)$, $\rho \mapsto -\rho$ the same scenario of the degeneration of the Lorenz dynamical chaos occurs with the change of the parameter ρ towards lower values. The only difference is that the right-hand equilibrium becomes globally stable. This scenario also takes place for the nonzero Λ, at least for $|\Lambda| < 3$. Asymmetric Lorenz attractors are shown in Fig. 3.4a, c. Note that for $\rho\Lambda > 0$ and small values of $|\Lambda|$, there exists a value of ρ for which the attractor loses its visual symmetry (see Fig. 3.4b).

5.2. The Feigenbaum attractor and the alternation. This type of chaotic attractors has been found for $\rho\Lambda > 0$ and sufficiently large values of Λ. In the numerical experiment, $|\Lambda| \geq 7$. The Feigenbaum attractor [9] is shown in Fig. 3.5a. If $\Lambda \leq -7$, then it is found that, during the increase of parameter ρ, the left equilibrium of system (3.11) loses stability when a stable limit cycle is born. Further, this cycle experiences a series of bifurcations of doubling of the period (pitchfork bifurcations). In some interval of values of parameter ρ, the chaotic attractor is born and has the attraction domain isolated from the attraction domains of other limit sets. With the further increase of parameter ρ, the attraction domain of the attractor intersects with the attraction domain of another chaotic limit set (it has not been studied, which one exactly, but probably this limit set is an "inheritance" of the asymmetric Lorenz attractor). As a result, a typical alternation [10] is observed (see Fig. 3.5b). The figure shows just one ejection of the trajectory from the attraction domain of the Feigenbaum attractor. Since system (3.11) is invariant with respect to transformation $(x, y, z) \mapsto (-x, -y, z)$, $\rho \mapsto -\rho$, $\Lambda \mapsto -\Lambda$, the same scenario occurs in the right-hand half-plane (x, z) for positive Λ and decreased values of parameter ρ.

For negative values of R and any values of other parameters, no chaotic attractors have been found in system (3.11).

Numerical experiment. As usual, such a numerical experiment is used to illustrate the analytical results and to specify some details.

The Poincaré mapping has been plotted using the secant hyperplane $\varphi = \text{const}$ mapped onto itself over period 2π, namely: $(\theta, \eta, \xi)_{\varphi=\varphi_0} \to (\overline{\theta}, \overline{\eta}, \overline{\xi})_{\varphi=\varphi_0+2\pi}$. For the rotational motions, secant $\varphi = \text{const}$ is global. For the sake of convenience, the mapping has been done for the following system that is equivalent to system (3.8):

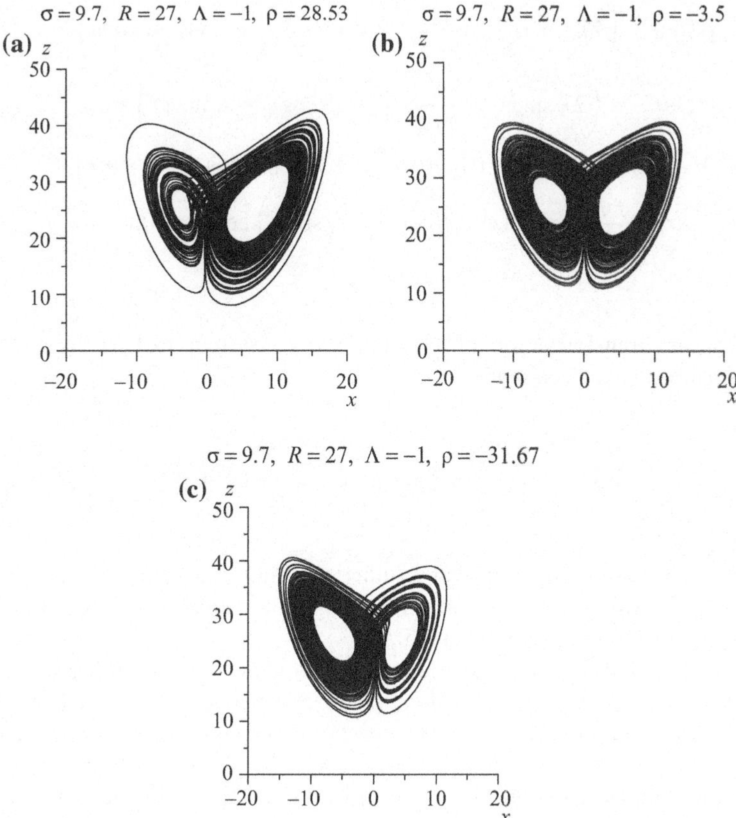

Fig. 3.4 Strange attractor. **a** asymmetric to the right; **b** symmetric; **c** asymmetric to the left

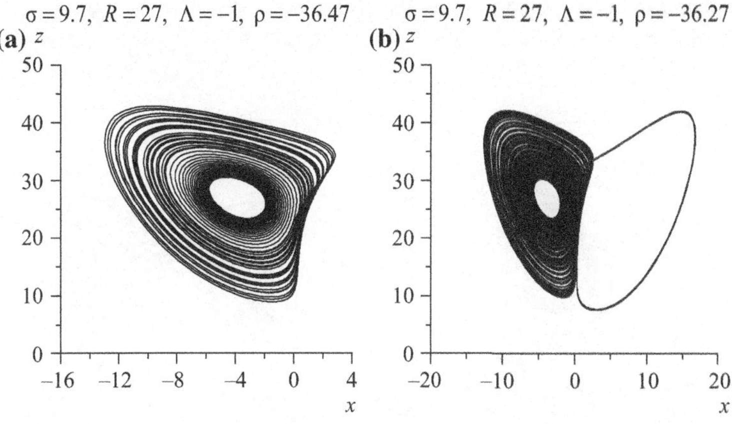

Fig. 3.5 The Feigenbaum attractor (**a**) and the alternation (**b**)

$$\dot{\theta} = \mu\left(\eta\xi + \left(2\lambda\sin\varphi + \frac{qr}{m\omega_0}\xi - 2h(\theta\cos\varphi - \eta\sin\varphi)\right)\cos\varphi\right),$$

$$\dot{\eta} = \mu\left(-\theta\xi - \left(2\lambda\sin\varphi + \frac{qr}{m\omega_0}\xi - 2h(\theta\cos\varphi - \eta\sin\varphi)\right)\sin\varphi\right),$$

$$\dot{\xi} = M_d - (\delta + qr^2)\omega_0 - \mu(\delta + qr^2)\xi + rq\omega_0(\theta\cos\varphi - \eta\sin\varphi)$$

$$+ c_1 r_1\left(\frac{rq}{m\omega_0} + \theta\sin\varphi + \eta\cos\varphi - r_1\sin\varphi\right)\cos\varphi - M_0\cos(\varphi + \varphi_0),$$

$$\dot{\varphi} = \omega_0 + \mu\xi. \tag{3.13}$$

During the transformation of system (3.8) to system (3.13), the following transformations have been made:

$$x = qr/m\omega_0 + \theta\sin\varphi + \eta\cos\varphi,$$
$$\dot{x} = (\theta\cos\varphi - \eta\sin\varphi)\omega_0, \quad \dot{\varphi} = \omega_0 + \mu\xi,$$

where $\mu = I^{-1}$.

To properly compare analytical and numerical results, the values of parameters of system (3.13) have been chosen such that their combinations would result in forms of attractors shown in Figs. 3.4 and 3.5.

Figure 3.6a shows an asymmetric Lorenz attractor in the Poincaré domain. A slight visual difference between this attractor and the one shown in Fig. 3.4a is caused by the transformation of coordinates. For these parameters, the parameters of the averaged system are as follows: $\sigma = 9.9$, $R = 27.39$, $\Lambda = -0.87$. Figure 3.6b shows a part of the rotational chaotic trajectory on the involute of the phase cylinder that corresponds to this chaotic attractor.

Figure 3.7a shows the Feigenbaum attractor originating from doublings of the period of the invariant torus. For chosen parameters of the original system, parameters of the averaged system (3.11) have the following values: $\sigma = 9.9$, $R = 27.39$,

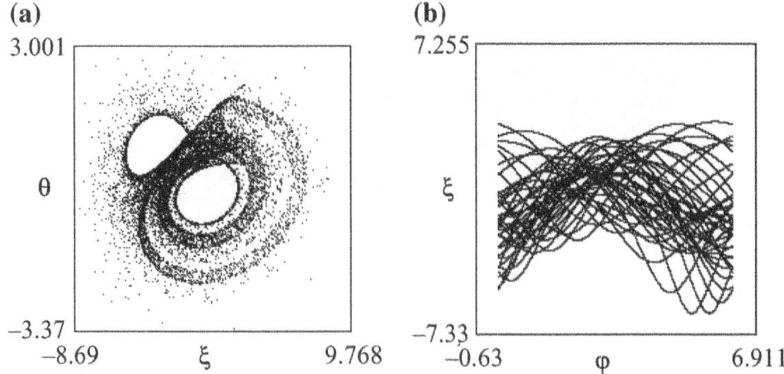

Fig. 3.6 Asymmetric (to the right) Lorenz attractor in the Poincare plane (**a**); a part of the chaotic trajectory on the projection on the involute of the phase cylinder (**b**)

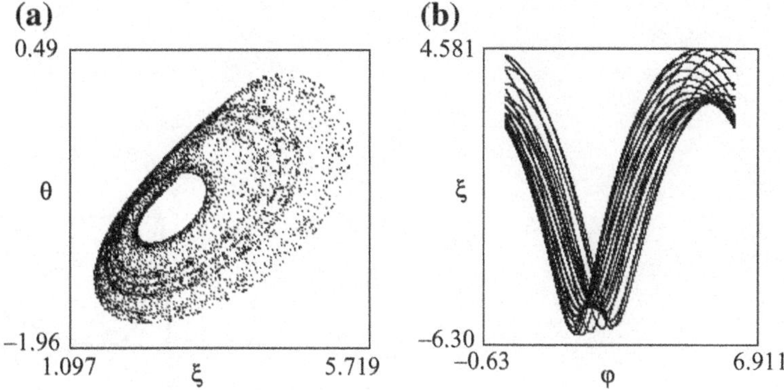

Fig. 3.7 The Feigenbaum attractor in the Poincare plane (**a**); a part of the chaotic trajectory on the projection on the involute of the phase cylinder (**b**)

$\Lambda = -7.061$. Figure 3.7b shows a part of the rotational chaotic trajectory on the involute of the phase cylinder.

To generalize the obtained information, let us study qualitative forms of the rotation characteristic in the resonance zone.

(1)   Suppose that $R \leq 0$. In this case, the dynamics of the system is quite simple. In the phase space of system (3.11), there exist only equilibria that correspond to limit cycles of system (3.8). These equilibria experience only one type of bifurcations: merging of equilibria with a following formation of saddle-focus and its further disappearance. In this case, depending on the parameters, the rotation characteristic has one or two hysteretic loops. Jumps in the frequency of the rotor rotations take place at extrema of the rotation characteristic. Figure 3.8

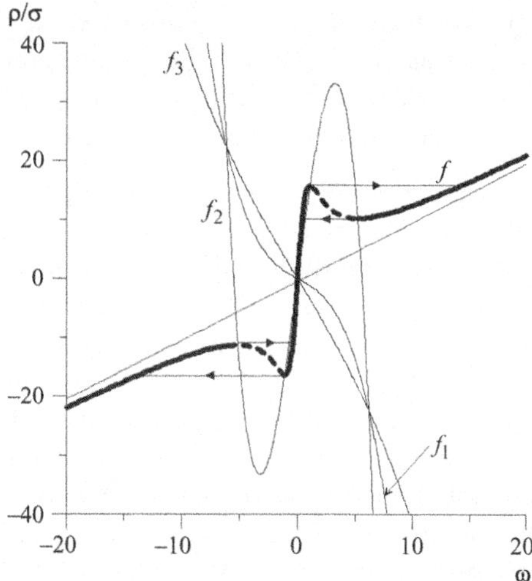

Fig. 3.8 Rotation characteristic in the resonance zone for $\sigma = 15$, $R = -30$, $\Lambda = -0.5$

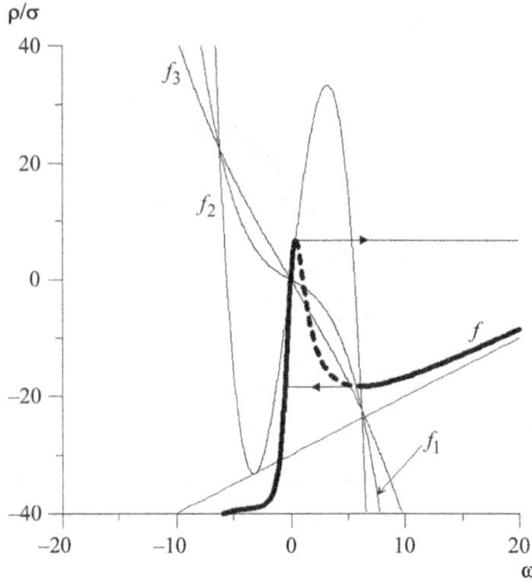

Fig. 3.9 Rotation characteristic in the resonance zone for $\sigma = 15$, $R = -30$, $\Lambda = -30$

shows a rotation characteristic with two loops (bold line). Dashed sections correspond to the unstable equilibria of saddle or saddle-focus type. Bold lines correspond to stable equilibria of system (3.11) [stable limit cycles of system (3.8)]. Thin lines correspond to auxiliary functions that define stability and types of equilibria [11]. One can say that a so-called "double" Sommerfeld effect takes place.

For decreased $|R|$, and for increased $|\Lambda|$, the left hysteretic loop disappears and only the right one remains (see Fig. 3.9). Thus, a classical Sommerfeld effect with one loop takes place. Due to the invariance of system (3.11) with respect to the transformation $(x, y, z) \mapsto (-x, -y, z)$, $\rho \mapsto -\rho$, $\Lambda \mapsto -\Lambda$, for increased positive values of Λ, the right hysteretic loop disappears, while the left one remains.

(2) Assume that $R > 0$. In this case, the dynamics of the system is more diverse during the variation of the system parameters. Accordingly, a set of the qualitative forms of the rotation characteristics is more diverse as well. In what follows, we consider only the cases of existence of chaotic attractors.

Suppose that $R^* < R < R_c$ (see Fig. 3.10a). In this case, in some interval of the values of parameter ρ/σ, there exist two stable equilibria and the Lorenz attractor. In the figures, this domain is shaded. In this area, depending on the initial conditions, either one of the two equilibria can occur [periodic motion of system (3.8)], as well as a strange attractor that corresponds to chaotic behavior of the instantaneous frequency of the rotator. Outside of this area, a regime or periodic beatings of the rotator takes place, which is realized for any initial conditions.

Figure 3.10b shows a rotation characteristic for $R > R_c$. In the shaded area, for any initial conditions, a strange attractor, whose properties change depending on the variation of parameter ρ/σ, is realized. In this case, the temporal average value $\langle x(t, t_0)\rangle_t$. also changes. Moreover, due to the strong dependence of trajectories of

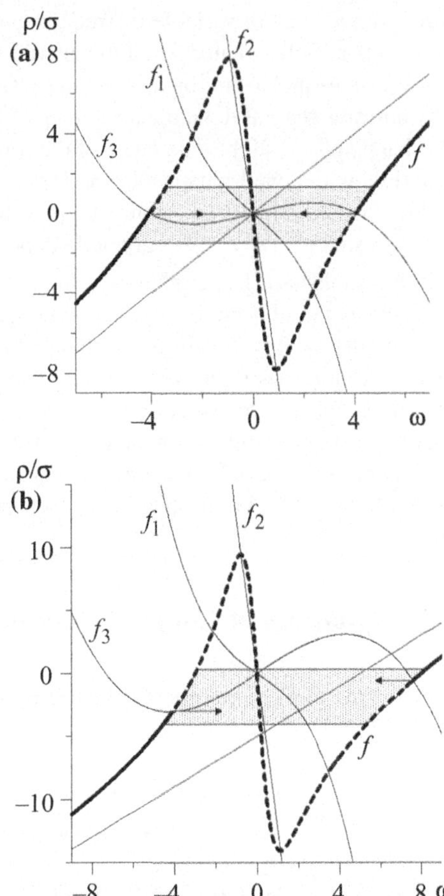

Fig. 3.10 Rotation characteristic in the resonance zone for $\sigma = 10$, $R = 17.5$, $\Lambda = 0$ (**a**); $\sigma = 10$, $R = 25$, $\Lambda = -5$ (**b**)

the attractor on the initial conditions, and due to the finiteness of the real interval of the averaging, this value will strongly depend on the initial moment of time t_0. Based on the aforesaid scenario, the following conclusions regarding the behavior of the rotation characteristic in the shaded area can be made:

(a) The rotation characteristic in the shaded domain is irreproducible: for a quasi-stationary increase of parameter ρ/σ (constant part of the motor's torque), one obtains one curve (branch), but for an opposite parameter variation (arbitrarily small), one obtains a completely different curve;

(b) The rotation characteristic in the shaded domain has an infinite number of branches that begins at the points of frequency jumps (from periodic to chaotic rotations) at the ends of bold lines (see Fig. 3.10).

To conclude this section, let us explain why an imbalance of the rotor can lead or cannot lead to a chaotic dynamical regime in the system under consideration. During the absence of a rotor imbalance, its equation represents an equation of the first order with respect to variable $\dot{\varphi}$ (frequency of rotations). This asynchronous motor represents a "full" rotator, i.e. a dynamical system with one degree of freedom. In this case, a formal averaging of system (3.8) results in a system of the second order, which means that all dynamical regimes of the model are regular. The same would also happen for a slight imbalance of the rotor. Due to this reason, dynamical chaos in system with a well-balanced rotor does not occur.

We would like to note the following. System parameter $R = \frac{2A\lambda}{(\delta+qr^2)h^2}\sin\varphi_0$, $A = \frac{M_0qr}{2\omega_0}$, crucially depends on the angle between the imbalance and the crankshaft. Depending on which, either a chaotic dynamics of vibrational mechanism would occur or, on the contrary, a stabilization of rotations would occur. In particular, if $R < 0$ and its absolute value is large enough, then a section of the rotation characteristic between hysteretic loops can be quite steep (see Fig. 3.8). Therefore, for a right selection of the position of imbalance and its mass, one can obtain a decent stabilization of motor rotations when external forces, including the random ones, would not dramatically affect the frequency of rotations. This is similar to the case of a master-slave synchronization of a rotator by external force.

3.3 Dynamics of Coupled Rotators

A model of coupled rotators is governed by a system of differential equations of the form [12, 13]

$$I\ddot{\varphi}_1 + \delta_1\dot{\varphi}_1 + \sigma_1\sin\varphi_1 = \gamma_1 + a_1\dot{\varphi}_2 + b_1\sin\varphi_2,$$
$$I\ddot{\varphi}_2 + \delta_2\dot{\varphi}_2 + \sigma_2\sin\varphi_2 = \gamma_2 + a_2\dot{\varphi}_1 + b_2\sin\varphi_1. \tag{3.14}$$

This system is defined in a toroidal phase space $G(\varphi_{1,2}, \dot{\varphi}_{1,2}) = T^2 \times R^2$ and it will be considered in the following parameter domain: $D = \{I > 0, \ \delta_{1,2} > 0, \ \gamma_{1,2} \geq 0, \ \delta_1\delta_2 - a_1a_2 > 0, \ \sigma_{1,2} \geq 0\}$.

We will be interested in solving the following problems of dynamics of system (3.14):

- determination of parameter domains that correspond to various topologically different motions of the system: oscillatory motions, rotary-oscillatory motions, and rotary motions of rotators;
- study of qualitative forms of trajectories and bifurcations of rotary motions;
- obtaining qualitative forms of the rotation characteristic of rotators.

1. *Dynamics of rotators and attracting sets of trajectories of system* (3.14).

It has been shown in [12] that, in the phase space $G(\varphi_{1,2}, \dot{\varphi}_{1,2})$ of system (3.14), there exists a toroidal absorbing region

$$G^*(\varphi_{1,2}, \dot{\varphi}_{1,2}) = (\forall\varphi_{1,2}, \ \omega_{1,2} - \omega_{1,2}^0 \leq \dot{\varphi}_{1,2} \leq \omega_{1,2} + \omega_{1,2}^0),$$

$$\omega_1 = \frac{\gamma_1\delta_2 + \gamma_2 a_1}{\delta_1\delta_2 - a_1 a_2}, \quad \omega_2 = \frac{\gamma_2\delta_1 + \gamma_1 a_2}{\delta_1\delta_2 - a_1 a_2},$$

such that all trajectories that enter it remain there forever, i.e. for $\tau \to \infty$. It means that for parameters that belong to domain D, all limit sets of trajectories of the system under consideration remain in the absorbing region G^* of the phase space and any solution $(\varphi_1(\tau, \tau_0), \varphi_2(\tau, \tau_0))$ has finite derivatives $\dot{\varphi}_1(\tau, \tau_0), \dot{\varphi}_2(\tau, \tau_0)$.

Let us subdivide the limit sets of phase trajectories within domain G^* into sub-classes:

K_{00} is the class of trajectories bounded by φ_1 and φ_2;
K_{0r} is the class of trajectories bounded by φ_1 and not bounded by φ_2;
K_{r0} is the class of trajectories bounded by φ_1 and not bounded by φ_2;
K_{rr} is the class of trajectories bounded by φ_1 and φ_2.

It is evident that $\langle\dot{\varphi}_1^*\rangle_\tau = 0, \langle\dot{\varphi}_2^*\rangle_\tau = 0$, if $L^* \in K_{00}$; $\langle\dot{\varphi}_1^*\rangle_\tau \neq 0, \langle\dot{\varphi}_2^*\rangle_\tau = 0$, if $L^* \in K_{r0}$; $\langle\dot{\varphi}_1^*\rangle_\tau = 0, \langle\dot{\varphi}_2^*\rangle_\tau \neq 0$, if $L^* \in K_{0r}$; $\langle\dot{\varphi}_1^*\rangle_\tau \neq 0, \langle\dot{\varphi}_2^*\rangle_\tau \neq 0$, if $L^* \in K_{rr}$. Determination of parameter domains that correspond to various classes of non-wandering trajectories will be carried out by the method of two-dimensional comparison systems [14].

For each equation of system (3.14), we introduce comparison systems of the form

$$A_1^\pm : I\ddot{\varphi}_1 + \delta_1\dot{\varphi}_1 + \sigma_1 \sin\varphi_1 = \delta_1\omega_1 + \mu_1^\pm,$$
$$A_2^\pm : I\ddot{\varphi}_2 + \delta_2\dot{\varphi}_2 + \sigma_2 \sin\varphi_2 = \delta_2\omega_2 + \mu_2^\pm, \tag{3.15}$$

where

$$\delta_{1,2}\omega_{1,2} + \mu_{1,2}^+ = \sup_{\forall(\varphi_{1,2},\dot{\varphi}_{1,2})\in G^*} \left(\gamma_{1,2} + a_{1,2}\dot{\varphi}_{1,2} + b_{1,2}\sin\varphi_{1,2}\right),$$

$$\delta_{1,2}\omega_{1,2} + \mu_{1,2}^- = \inf_{\forall(\varphi_{1,2},\dot{\varphi}_{1,2})\in G^*} \left(\gamma_{1,2} + a_{1,2}\dot{\varphi}_{1,2} + b_{1,2}\sin\varphi_{1,2}\right),$$

$$\mu_{1,2}^\pm = \pm\left(|a_{1,2}|\omega_{1,2}^0 + |b_{1,2}|\right).$$

Trajectories of the comparison systems define surfaces that have no contact with the vector field of (3.14) in the phase space $G(\varphi_{1,2}, \dot{\varphi}_{1,2})$. This can be proven by considering the rotation of the vector field of the system on the trajectories of the comparison systems. It means that if both systems A_k^+ and A_k^- have topologically similar structures of trajectories on the involutes of cylinders $(\varphi_k, \dot{\varphi}_k)$, then projections of limit sets $K_{(.)}$ onto these involutes turn out to lie between special manifolds of comparison systems. Figure 3.11 shows separatrices of saddles and limit cycles of comparison systems. Arrows indicate direction of the component of the vector field of system (3.14) along these trajectories.

Let us introduce the following notations: $D_{(0)k}^\pm$, $D_{(0r)k}^\pm$, $D_{(r)k}^\pm$, $k = 1, 2$ for the parameter domains, for which both comparison systems $A_k^\pm$ (separately for each index k) have globally stable equilibrium, and globally stable equilibrium and limit

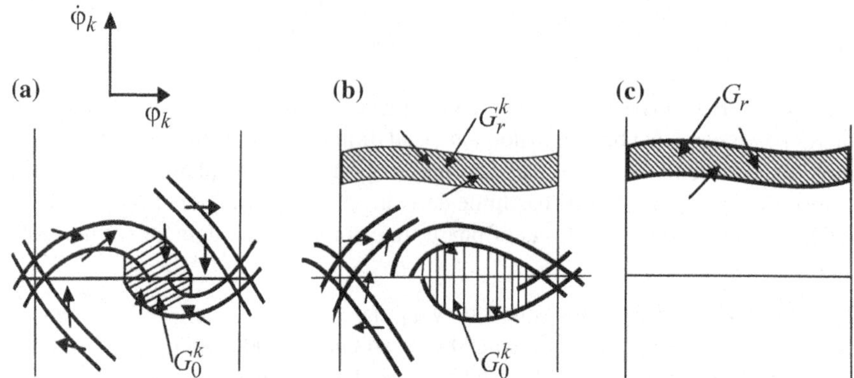

Fig. 3.11 Qualitative forms of trajectories of comparison systems and orientation of projection of the vector field of system (3.14) on their special trajectories

cycle, respectively (see Fig. 3.11, as well as Sect. 1.1). For parameter planes $\left(\lambda_k, \gamma_k^{\pm}\right)$, where $\lambda_k = \delta_k(I\sigma_k)^{-1/2}$, $\gamma_k^{\pm} = \frac{\delta_k\omega_k + \mu_k^{\pm}}{\sigma_k}$, these domains are separated by line $\gamma_k^{\pm} = 1$ (bifurcation of equilibrium) and Tricomi curve $\lambda_k = \lambda_k^*\left(\gamma_k^{\pm}\right)$ (bifurcation of the separatrix loop). Bifurcation diagrams of comparison systems are shown in Fig. 3.12. The hatching shows the projections of the absorbing regions for various classes of trajectories.

It is evident that for parameter domains $D_{(.)}^k = D_{(.)k}^+ \cap D_{(.)k}^-$, comparison systems $A_k^{\pm}$ have similar structures of trajectories on involutes $(\varphi_k, \dot{\varphi}_k)$. From the aforesaid case, we can make the following statements regarding parameter domains of existence of the aforementioned classes of limit sets of trajectories of system (3.14).

For parameter domain $D_{00} = D_0^1 \times D_0^2$, there exists a limit set of trajectories $K_{00} \in G_0^1 \times G_0^2$, which is realized for any initial condition; for parameters $D_{0r} = D_0^1 \times D_r^2$, there exists a globally stable limit set $K_{0r} \in G_0^1 \times G_r^2$: from any initial condition, the system reaches the state when the first rotator experiences oscillatory motion, while the second one is rotating (we are not yet talking about specific forms of movements); for parameter domain $D_{r0} = D_r^1 \times D_0^2$, there exists a globally stable

Fig. 3.12 Combined bifurcation diagram for comparison systems

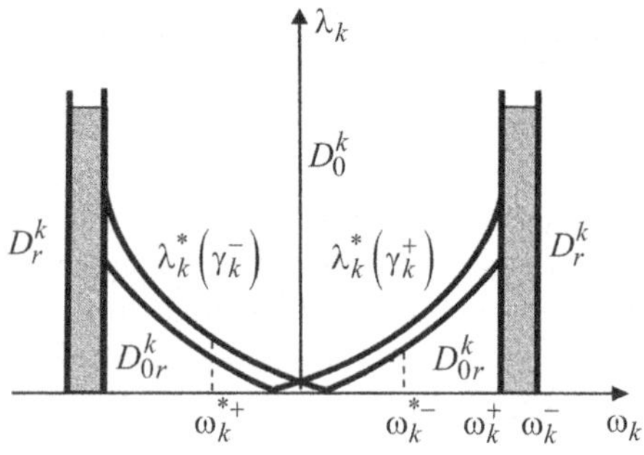

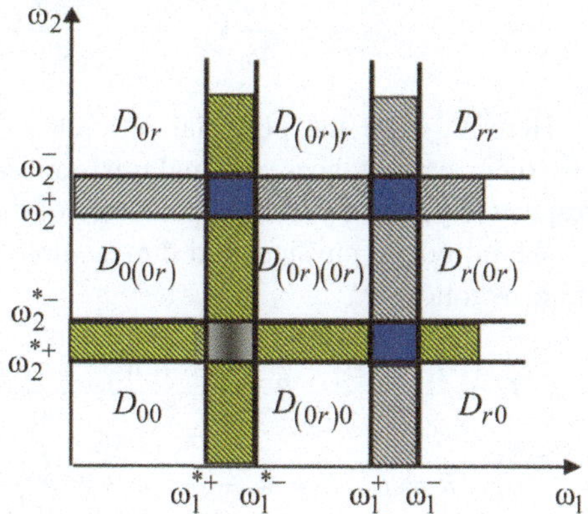

Fig. 3.13 Decomposition of parameter plane into domains that correspond to different classes of trajectories

limit set $K_{r0} \in G_r^1 \times G_0^2$; for parameter domain $D_{rr} = D_r^1 \times D_r^2$, there exists a globally stable limit set $K_{rr} \in G_r^1 \times G_r^2$. Outside of those parameter domains, there exist populations of sets of trajectories of different types that are realized depending on the initial conditions. A mutual distribution of parameter domains is shown in Fig. 3.13.

Let us introduce another form of system of coupled rotators, for which we consider a physical example [13]. Figure 3.14 shows an electric scheme of two superconductive junctions with a capacitive coupling.

In physical variables and parameters, the Kirchhoff rules and the Josephson relation result in the following dynamical model of the system under consideration:

$$C_1 \frac{dV_1}{dt} + R_1^{-1} V_1 + I_{c1} \sin \varphi_1 = I - C_0 \frac{d(V_1 + V_2)}{dt},$$

$$C_2 \frac{dV_2}{dt} + R_2^{-1} V_2 + I_{c2} \sin \varphi_2 = I - C_0 \frac{d(V_1 + V_2)}{dt},$$

Fig. 3.14 Electric scheme of two superconductive junctions with capacitive coupling

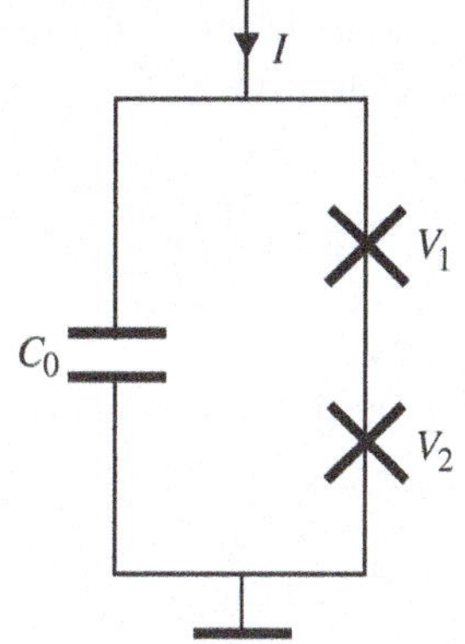

$$\frac{d\varphi_1}{dt} = \frac{2\pi}{\Phi_0} V_1, \; \frac{d\varphi_2}{dt} = \frac{2\pi}{\Phi_0} V_2. \tag{3.16}$$

Here $V_{1,2}$ are the potential differences of superconductors; $C_{1,2}$, $R_{1,2}$, $I_{c1,c2}$ are their capacitances, normal resistances, and values of critical overcurrent, respectively; and C_0 is the capacitance of the coupling.

We introduce physical and dimensionless parameters, as well as dimensionless time, as follows:

$$I_c = \frac{I_{c1} + I_{c2}}{2}, \; R = \frac{R_1 + R_2}{2}, \; \frac{2\pi R I_c}{\Phi_0} = \Omega, \; \Omega t = \tau, \; C_{1,2}\Omega R = c_{1,2},$$

$$C_0 \Omega R = c_0, \; r_{1,2} = \frac{R_{1,2}}{R}, \; \nu_{1,2} = \frac{I_{c1,c2}}{I_c}, \; i = \frac{I}{I_c}.$$

Instead of system (3.16), we obtain a dimensionless system of the form

$$c_1 \ddot{\varphi}_1 + r_1^{-1}\dot{\varphi}_1 + \nu_1 \sin \varphi_1 = i - c_0(\ddot{\varphi}_1 + \ddot{\varphi}_2)$$
$$c_2 \ddot{\varphi}_2 + r_2^{-1}\dot{\varphi}_2 + \nu_2 \sin \varphi_2 = i - c_0(\ddot{\varphi}_1 + \ddot{\varphi}_2). \tag{3.17}$$

By solving system (3.17) with respect to higher-order derivatives, we obtain system (3.14). Parameters of these two systems are related to each other as follows:

$$I = \frac{c_1 c_2 + c_0(c_1 + c_2)}{c_0}, \; \delta_{1,2} = \left(1 + \frac{c_{2,1}}{c_0}\right)\frac{1}{r_{1,2}}, \; \sigma_{1,2} = \left(1 + \frac{c_{2,1}}{c_0}\right)\nu_{1,2},$$

$$\gamma_{1,2} = \frac{c_{2,1}}{c_0} i, \; a_{1,2} = \frac{1}{r_{2,1}}, \; b_{1,2} = \nu_{2,1}, \; \omega_{1,2} = r_{1,2} i.$$

In what follows, in addition to system (3.14), we consider an equivalent system (3.17). The following statement answers the question on the composition of the limit set of trajectories K_{00}.

System (3.17), as well as system (3.14), have no closed limited phase trajectories (limit cycles of oscillatory type). This property is proven by means of the periodic Lyapunov function

$$V = \frac{1}{2}\sum_{k=1}^{2} c_k c_0^{-1} \dot{\varphi}_k^2 + c_0^{-1}\sum_{k=1}^{2} \int_{\varphi_{0k}}^{\varphi_k} (\nu_k \sin \varphi_k - i)d\varphi_k + \frac{1}{2}(\dot{\varphi}_1 + \dot{\varphi}_2)^2,$$

The derivative of which along the vector field of system (3.17) has the form

$$\dot{V} = -\frac{1}{r_1 c_o}\dot{\varphi}_1^2 - \frac{1}{r_2 c_o}\dot{\varphi}_2^2,$$

i.e. the derivative is non-positive in the whole phase space $G\left(\varphi_{1,2}, \dot{\varphi}_{1,2}\right)$. This means that limit trajectories of limit set K_{00} are (only) the equilibria: stable equilibrium $O_1(\varphi_1, \varphi_2, \dot{\varphi}_1, \dot{\varphi}_2) = O_1\left(\arcsin \frac{i}{v_1}, \arcsin \frac{i}{v_2}, 0, 0\right)$ (knot or focus) and saddle-type equilibria: $O_2\left(\pi - \arcsin \frac{i}{v_1}, \arcsin \frac{i}{v_2}, 0, 0\right)$, $O_3\left(\arcsin \frac{i}{v_1}, \pi - \arcsin \frac{i}{v_2}, 0, 0\right)$, $O_4\left(\pi - \arcsin \frac{i}{v_1}, \pi - \arcsin \frac{i}{v_2}, 0, 0\right)$. The only bifurcation of equilibria is the fusion of the entire four equilibria that is followed by the formation of a complex equilibrium that disappears at $i = \min(v_1, v_2) + 0 = i_0 + 0$.

2. *Structures of trajectories and bifurcations of rotary motions of rotators.*

Let us consider a qualitative structure of the set of rotary motions K_{rr}. We will select parameters of the system from those domains, for which rotary motions of both rotators occur (see Fig. 3.13). This can take place for initial conditions for a part of the phase space if parameters are chosen from domains $D_{(0r)(0r)}$, $D_{(0r)r}$, $D_{r(0r)}$, or for the whole phase space if parameters are chosen from domain D_{rr}.

Let us consider asymptotic case $I \gg 1$. Using a change of the variables (see Appendix I) of the form

$$\dot{\varphi}_1 = \omega_1 + \mu \Phi_1(\varphi_1, \varphi_2, x_1), \ \dot{\varphi}_2 = \omega_2 + \mu \Phi_2(\varphi_1, \varphi_2, x_2),$$

$$\Phi_1 = \frac{\sigma_1}{\omega_1} \cos \varphi_1 - \frac{b_1}{\omega_2} \cos \varphi_2 + x_1, \ \Phi_2 = \frac{\sigma_2}{\omega_2} \cos \varphi_2 - \frac{b_2}{\omega_1} \cos \varphi_1 + x_2,$$

$$\mu = I^{-1}$$

we reduce system (3.14) to an equivalent system with fast-spinning phases $\varphi_{1,2}$. In the zone of the main resonance of $\omega_1 \approx \omega_2$, the averaged system has the form

$$\dot{x}_1 = \mu(-\delta_1 x_1 + a_1 x_2 - B_1 \sin \eta),$$
$$\dot{x}_2 = \mu(-\delta_2 x_2 + a_2 x_1 + B_2 \sin \eta),$$
$$\dot{\eta} = \mu(\Delta + x_1 - x_2). \tag{3.18}$$

Here $\eta = \varphi_1 - \varphi_2$ is the phase mistuning, $\omega_1 - \omega_2 = \mu \Delta$ is the frequency mistuning, $B_1 = \frac{b_1(\sigma_1 + b_2)}{2\omega_1 \omega_2}$, $B_2 = \frac{b_2(\sigma_2 + b_1)}{2\omega_1 \omega_2}$.

Note that the condition of closeness of normal frequencies of rotators $\omega_{1,2}$ means that, first, parameters of the system are chosen from a small vicinity of the diagonal line of plane (ω_1, ω_2), and, second, closeness of parameters $r_{1,2} : \mu \Delta = (r_1 - r_2)i$.

In turn, Eq. (3.18) for a change of time $\mu \Omega_0 \tau = \tau_n$, $\Omega_0 = \left(\frac{B_1(\delta_2 - a_2) + B_2(\delta_1 - a_1)}{\delta_1 + \delta_2}\right)^{\frac{1}{2}}$ are reduced to an already known equation of the third order (see Sect. 2.1):

$$\beta \dddot{\eta} + \ddot{\eta} + \alpha \cos \eta \dot{\eta} + \lambda' \dot{\eta} + \sin \eta = \gamma', \tag{3.19}$$

where $\beta = \Omega_0/(\delta_1 + \delta_2)$, $\alpha = (B_1 + B_2)/\Omega_0(\delta_1 + \delta_2)$, $\lambda^r = \frac{\delta_1\delta_2 - a_1 a_2}{\Omega_0(\delta_1 + \delta_2)}$, $\gamma^r = \frac{\delta_1\delta_2 - a_1 a_2}{\Omega_0^2(\delta_1 + \delta_2)}\Delta$. Qualitative structures of trajectories of this equation are already known (see Sect. 2.1), while their interpretation onto system (3.14) is mentioned above.

Note that, for identical symmetrically coupled rotators, system (3.18) has a unique globally stable integral manifold $x_1 = -x_2$. This manifold is decomposed into trajectories by trajectories of the equation of a free pendulum:

$$\ddot{\eta} + (\delta + a)\dot{\eta} + 2B \sin \eta = 0. \tag{3.20}$$

During the transformation of system (3.18) to system (3.20), a transformation of time has been made $\mu\tau = \tau_n$. In turn, this equation has equilibrium $O(\eta, \dot{\eta}) = O(0, 0)$, which is globally stable, i.e. the dynamics of identical slightly nonlinear rotors in the parameter domain D is quite simple: for any initial condition over the course of time, a synchronization of rotary motions is observed in the coupled system.

3. *Rotation characteristics of rotators.*

By studying equilibria of system (3.17), we solve the problem of zero steps of the rotation characteristics of rotators that correspond to the superconducting branches of the current-voltage characteristics of the superconductive junctions in the circuit shown in Fig. 3.14. Zero steps exist and are stable in the interval $0 \le i < i_0$.

To describe the resistive part of the current-voltage characteristic of superconductive junctions, let us turn to system (3.18) and to Eq. (3.19). Equation (3.19) has two equilibria: $O_1(\eta, \dot{\eta}, \ddot{\eta}) = O_1(\arcsin \gamma^r, 0, 0)$ (knot or focus) and $O_2(\pi - \arcsin \gamma^r, 0, 0)$ (saddle). Equilibrium O_1 is stable in the whole area of existence (see Sect. 2.1) and corresponds to the limit cycle of system (3.14), i.e. to the regime of stable mutual synchronization of rotators. A holding range (range of existence) of synchronization is defined by a system of inequalities: $|\gamma^r| \le 1, i > i^*$. The first one corresponds to existence of equilibrium of Eq. (3.19), while the second one defines the domain of rotations of rotators. i^* defines the border that separates the domain of rotations from the domain of global stability of equilibrium of system (3.17). The exact value i^* is determined experimentally. Also, the following estimation is valid: $i^* = \max\left(r_1^{-1}\omega_1^{*-}, r_2^{-1}\omega_2^{*-}\right)$. In the expanded form, the region of existence of synchronization is determined by a double inequality of the form

$$i^* < i < i_*, \; i_* = \left(\frac{c_0 v_1 v_2}{|r_1 - r_2|(c_1 + c_2)(c_1 c_2 + c_0 c_1 + c_0 c_2)}\right)^{\frac{1}{3}}.$$

In the regime of synchronization, $\Omega_1 = \Omega_2 = \frac{\omega_1 + \omega_2}{2} = \frac{r_1 + r_2}{2} i$ is the mean value of partial frequencies, which is quite natural.

To obtain the full picture of the rotation characteristic, it is necessary to add branches that correspond to the regimes of beatings of rotators to the synchronization branch.

From the boundedness of derivatives $\dot{\varphi}_{1,2}$, it follows that $\left\langle \dddot{\varphi}_1^* \right\rangle_\tau - \left\langle \dddot{\varphi}_2^* \right\rangle_\tau = \left\langle \dddot{\eta}^* \right\rangle_\tau = 0$. In this case, by using a change of variables and system (3.18), we obtain:

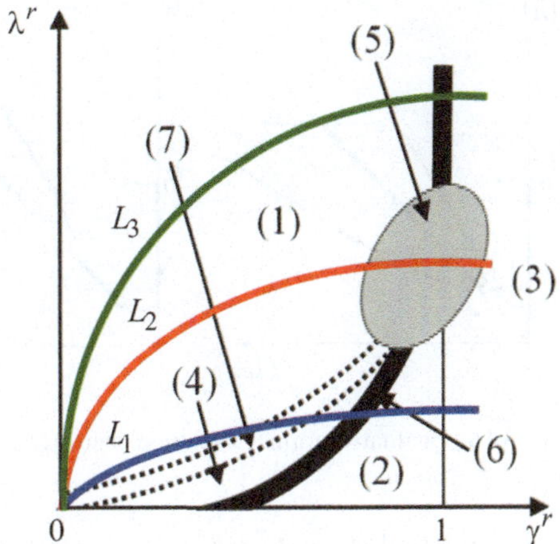

Fig. 3.15 Bifurcation diagram of parameters of Eq. (3.19)

$$\Omega_{1,2} = \omega_{1,2} + d^{-1}\left((\delta_{2,1} + a_{1,2})\left(\langle \dot{\eta}^* \rangle_\tau - \mu\Delta\right) + \mu(B_1 + B_2)\langle \sin \eta^* \rangle_\tau\right),$$

$$d = \delta_2 - \delta_1 + a_1 - a_2. \tag{3.21}$$

It follows from the properties of Eq. (3.19) that for increased parameter γ^r (increased parameter Δ, and, therefore, increased parameter i), $\langle \dot{\eta}^* \rangle_\tau \rightarrow \mu\Delta$, $\langle \sin \eta^* \rangle_\tau \rightarrow 0$. Therefore, $\Omega_{1,2} \rightarrow \omega_{1,2}$, i.e. lines $i = r_{1,2}^{-1}\Omega_{1,2}$ represent asymptotes for the branches of beatings. If we assume that $r_1 > r_2$, then $\gamma^r > 0$, $\langle \dot{\eta}^* \rangle_\tau > 0$, and, therefore, $\Omega_1 - \Omega_2 = \langle \dot{\eta}^* \rangle_\tau > 0$, $\Omega_1 > \Omega_2$. The branches of the partial frequencies of the rotation characteristics that correspond to the beatings are located on the opposite sides of the synchronization branch. We also note that when the parameter i is varied, then parameters $\gamma^r \sim i^3$ and $\lambda^r \sim i$ vary along line $\gamma^r = \alpha(\lambda^r)^3$, where $\alpha > 0$.

Figure 3.15 shows one of the bifurcation diagrams in the plane (λ^r, γ^r) and lines $L_{1,2,3}\left(\gamma^r = \alpha_{1,2,3}(\lambda^r)^3\right)$, which correspond to qualitatively different cases.

Let us remind the reader that for the parameters from region (1) in the phase space of system (3.14), there exists a stable (synchronization) and unstable limit cycles lying on the resonance torus T^2 For the parameter domain (4), resonance limit cycles, as well as stable two-dimensional torus T_1^2 and unstable two-dimensional torus T_2^2 (beatings), exist in the phase space. For the parameter domain (2), there exist aforementioned limit cycles (lying on the torus; T^2 gets destroyed in the parameter domain (6)), as well as invariant torus T_1^2 (beatings). For the parameter domain (3), there exists torus T_1^2 which is globally stable if parameters are chosen in the parameter domain D_{rr}. It is possible that for certain variations of parameters, subharmonic resonances that correspond to multiple synchronization may originate and disappear on the aforementioned tori [5, 15].

Figure 3.16 shows qualitative forms of rotation characteristics of both rotators for various parameters of the system.

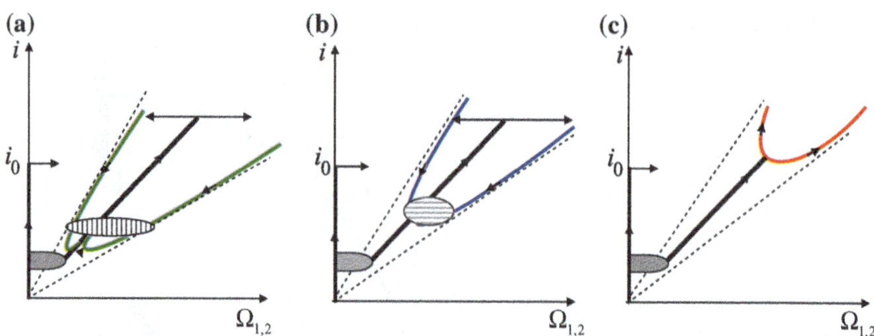

Fig. 3.16 Qualitative forms of rotation characteristics of rotators

A. Suppose that parameter i is quasi-statically increased from zero, and parameters λ^r, γ^r are varied (increased) along line L_1. In this case, rotators that remain in the equilibrium for $i = 0$ remain there until equilibrium disappears for $i = i_0 + 0$. This situation defines a zero branch of the rotation characteristic of both rotators. For $i = i_0 + 0$, both rotators suddenly jump into the regime of rotations. Depending on the initial condition, a simple periodic regime of the synchronization or a regime of quasi-periodic beatings can be realized. In the synchronization regime, rotation characteristics have a common branch, shown in Fig. 3.16 by a straight line (within the first approximation by a small parameter). Then the synchronization regime, if it has been realized, is abruptly (the bifurcation of the fusion of stable and unstable resonant limit cycles and their disappearance) replaced by the regime of beatings. The branches of rotation characteristic that correspond to the beatings are located on opposite sides of the synchronization branch. Let us now assume that parameter i is decreased and motions of parameters along line L_1 occur in the parameter domain (3) (see Fig. 3.15). In this case, each rotator remains in the regime of stable beatings until torus T_1^2 disappears due to the merging with unstable torus T_2^2 in domain (7). In the point of disappearance of T_1^2, rotators jump to the regime of synchronization. The upper shaded area, shown in Fig. 3.16a, corresponds to parameter domain (6) shown in Fig. 3.15: here, a destruction of the resonance torus and formation of homoclinic structures related to the homoclinic curve of the saddle resonance limit cycle occur. On the rotation characteristic, this domain is quite narrow, and so is domain (6) ($\sim\mu$), and the chaotic limit set is not an attractor itself. The lower shaded area shown in Fig. 3.16 is related to an unexplored area of parameters. One can state only one thing: the transition of rotators to equilibrium cannot be related to any limiting sets of trajectories of oscillatory type (regular or chaotic).

B. This case differs from the previous one (see Fig. 3.16b): the system returns from the regime of beatings to the regime of synchronization through the chaotization of the rotator's rotations. As it is known from Sect. 2.1, the parameter domain (5) (see Fig. 3.15) corresponds to chaotic attractors that are associated with the bifurcation of the destruction of the resonant torus T^2 during the transition of parameters from domain (1) to domain (5); with the bifurcation of the destruction of torus T_1^2 during the transition of parameters from domain(3) to domain (5); and with bifurcation of a homoclinic trajectory of saddle resonance limit cycle, as well as with pitchfork bifurcations of a two-dimensional torus. Chaotic attractors are illustrated by the numerical experiment in Sect. 2.1. A scattering of the rotation characteristic of rotators occurs in shaded domains as in Fig. 3.16b.

C. This case (see Fig. 3.16c) differs from the previous ones by a soft origination of the regime of quasi-periodic beatings at the moment when the system exits the regime of synchronization: the branches of the rotation characteristic that correspond to the regime of beatings adjoin to a straight line that corresponds to the synchronization.

For a system with Josephson junctions (see Fig. 3.14), rotation characteristics shown in Fig. 3.16 represent their volt-ampere characteristics.

References

1. Belykh, V.N., Verichev, N.N.: Dynamics of a rotator coupled with an oscillator. Radiophys. Quantum Electron. **31**(8), 657 (1988)
2. Andronov, A.A., Vitt, A.A., Khaikin, S.E.: Theory of Oscillators. Dover Publications, New York (1987)
3. Alifov, A.A., Frolov K.V.: Interaction of Non-linear Oscillatory Systems with Energy Sources. Taylor & Francis (1990)
4. Butenin, N.V., Neimark, Yu.I., Fufaev, N.A.: Introduction to the Theory of Nonlinear. Oscillations, Nauka, Moscow, 1976 (in Russian)
5. Blekhman, I.I.: Vibrations in Engineering, V.2. Vibrations of Nonlinear Mechanical Systems. Mashinostroenie, Moscow (1979) (in Russian)
6. Verichev, N.N., Verichev, S.N., Erofeyev, V.I.: Chaotic dynamics of simple vibrational systems. Int. J. Sound Vib. **310**, 755–767 (2008)
7. Verichev, N.N.: Study of systems with Josephson junctions by the method of a spinning phase. Radiotechnika i Elektronika **31**(11), 2267–2274 (1986). (in Russian)
8. Afraimovich, V.S., Bykov, V.V., Shilnikov, L.P.: On structurally stable attracting limit sets of Lorenz-attractor type. Trudy Mosk. Mat. Obshch, **46,** 153–216 (1983); English transl Trans. Moscow Math. Soc. **2** (1983)
9. Feigenbaum, M.J.: Quantitative universaliti for a class nonlinear trasformations. J. Stat. Phys. **19**(1), 25–52 (1978)
10. Arnold, V.I., Gamkrelidze, R.V.: Progress in Science and Technology. Current Problems in Mathematics. Fundamental Directions 3 (Dynamical Systems III). VINITI AN SSSR, Moscow (1985)
11. Mintz, R.M.: Study of the trajectories of a system of three differential equations at the infinity. Collection of papers in memory of A.A. Andronov, M., izd. AN SSSR, 1955 (in Russian)
12. Belykh, V.N., Verichev, N.N.: On dynamics of coupled rotators. Izv. Vuzov Radiofizika **XXXI**(6), 688–697 (1988) (in Russian)
13. Belykh, V.N., Verichev, N.N.: On the complex dynamics of an autonomous system with Josephson junctions. Radiotekhnika i elektronika. **31**(1), 140–147 (1987) (in Russian)
14. Akimov, N., Belyustina, L.N., Belykh, V.N., et al.: Phase Synchronization Systems (Radio Svyaz', Moscow) (1982) (Russian)
15. Lorenz, E.: Deterministic nonperiodic flow. J. Atmos. Sci. **20**, 130–141 (1963)

Chapter 4
Vibration of Shafts

4.1 Nonlinear Resonance of Bending Vibrations of a Flexible Rotor in a System with an Energy Source of Limited Power

Experimental studies and exploitation of rotating machinery (turbines, centrifugal pumps, etc.) show that the energy of lateral vibrations might be comparable with the energy of a power supply. This can happen during the startup process, when system has to bypass critical (resonant) speeds, as well as during the shutdown process. It can be the case that an operating speed is close to one of the resonant frequencies. Finally, a shift of resonance frequencies towards the operating speed can also occur during changes of oscillatory properties of the system during exploitation (wear of bearings, change of the mass e.g. due to sedimentation of various deposits, etc.). In all of these cases, the design of rotary systems and study of their dynamics cannot be properly done if dynamical properties of the energy source are omitted, i.e. one has to consider a coupled system of *"flexible rotor – energy source"*.

The effect of "jamming" of the rotational speed of the actuating mechanism in the vicinity of the resonant frequency of the oscillatory system, known as the Sommerfeld effect [1–4], also takes place for rotating shafts. This effect explains the increase of shaft vibrations in the resonance parameter domain, at which a significant part of the energy of the source passes into the energy of bending vibrations.

The problem of lateral vibrations of unbalanced shafts (flexible rotors) in a system with an energy source of limited power is not a new, but still an actual problem. There exists a large number of engineering publications on this topic. The analytical study, proposed in [5, 6], has not been completed due to the high dimension and complexity of the system. Below, we present a more detailed study that provides explicit forms of resonance characteristics of the flexible rotor and rotation characteristics of the actuating mechanism (energy source).

© Springer Nature Switzerland AG 2020

N. Verichev et al., *Chaos, Synchronization and Structures in Dynamics of Systems with Cylindrical Phase Space*, Understanding Complex Systems, https://doi.org/10.1007/978-3-030-36103-7_4

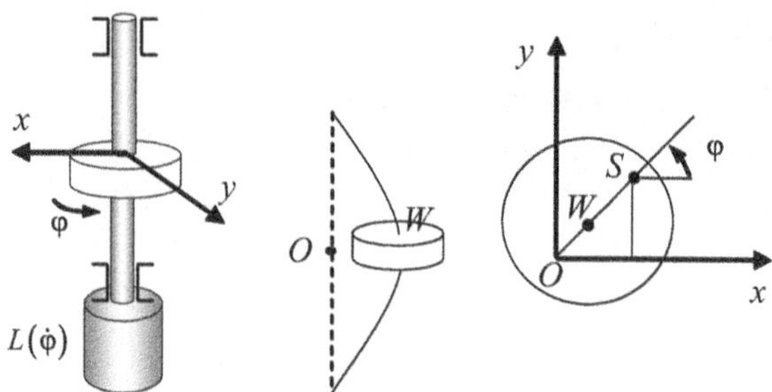

Fig. 4.1 Mechanical model of the system *"flexible rotor – energy source"*. Left figure: shaft with imbalanced disk; middle figure: shaft deflection; right figure: disk plane (view from above) showing coordinate plane and eccentricity

Consider a simple one-disk model of the flexible rotor, see Fig. 4.1. We will assume that the mass of the shaft distributed along its length, and its other equipment, are reduced to the disk. Such an imbalanced disk is placed in the center of a weightless and torsionless shaft. Here e is the disk eccentricity, point W designates the geometrical center of the disk, point S designates its gravity center, and point O designates the axis of the unperturbed shaft (see Fig. 4.1).

In the dimensional variables and parameters, the dynamics of this system is governed by the equations of the form [5, 6]

$$m\ddot{x} + \varepsilon\dot{x} + cx + k(\dot{x} + \dot{\varphi}y) = ce\cos\varphi,$$

$$m\ddot{y} + \varepsilon\dot{y} + cy + k(\dot{y} - \dot{\varphi}x) = ce\sin\varphi,$$

$$I\ddot{\varphi} = L(\dot{\varphi}) - q\dot{\varphi} - ce(x\sin\varphi - y\cos\varphi) - k(\dot{x}y - \dot{y}x) - k\dot{\varphi}(x^2 + y^2). \quad (4.1)$$

Here x, y are the coordinates of the disk's gravity center in the immovable coordinate system perpendicular to the unperturbed shaft axis, whose origin belongs to the shaft; ε, k are the coefficients of external and internal damping (for lateral vibrations); c is the stiffness of the shaft at the point of the disk's placement; m is the mass of the disk; I is the moment of inertia of the motor; φ is the angle of rotation; and $L(\dot{\varphi})$ is the torque of the motor including the moment of internal forces of resistance to the motion of the rotor. We assume that the torque is a linear function of the form $L(\dot{\varphi}) = T - q\dot{\varphi}$, where T is a constant component, and $q\dot{\varphi}$ is a momentum of the resistance forces against the motion of the rotor (e.g. damping inside the bearings) [4]. For the dimensionless variables and parameters, system (4.1) takes the form

$$\ddot{x} + \frac{\varepsilon\omega_0}{c}\dot{x} + x + \frac{k\omega_0}{c}(\dot{x} + \dot{\varphi}y) = \frac{e}{A_0}\cos\varphi,$$

$$\ddot{y} + \frac{\varepsilon\omega_0}{c}\dot{y} + y + \frac{k\omega_0}{c}(\dot{y} - \dot{\varphi}x) = \frac{e}{A_0}\sin\varphi,$$

$$\frac{I\omega_0^2}{T_0}\ddot{\varphi} = \frac{T}{T_0} - \frac{q\omega_0}{T_0}\dot{\varphi} - \frac{ceA_0}{T_0}(x\sin\varphi - y\cos\varphi)$$

$$- \frac{k\omega_0 A_0^2}{T_0}(\dot{x}y - \dot{y}x + \dot{\varphi}(x^2 + y^2)). \quad (4.2)$$

To pass from system (4.1) to system (4.2), we have introduced dimensionless time $\omega_0 t = \tau$, $\omega_0^2 = c/m$, as well as scaled variables x, y by constant A_0 and scaled constant moments by constant T_0. The values of the scaled variables are not of fundamental importance. The dimensionless variables keep the same notations as the dimensional ones. We introduce new parameters

$$c/m = \omega_0^2,\, q\omega_0/T_0 = \gamma,\, \left(I\omega_0^2/T_0\right)^{-1} = \mu,\, \varepsilon\,\omega_0/c = \mu h,\, k\omega_0/c = \mu h_1,$$
$$e\big/A_0 = \mu\,\nu,\, ce A_0/T_0 = \mu\,\lambda,\, k\omega_0 A_0^2/T_0 = \mu\,\chi,\, (T - q\omega_0)\big/T_0 = \mu\,\Delta.$$

In this case, it is not necessary to perform "complex" transformations of the system that have been outlined above. We introduce a new variable of the form $\dot{\varphi} = 1 + \mu\xi$ (for the new time, the frequency of oscillators equals 1), after which system (4.2) takes the form

$$
\begin{aligned}
\dot{x} &= x_1,\\
\dot{x}_1 &= -x + \mu F_1,\\
\dot{y} &= y_1,\\
\dot{y}_1 &= -y + \mu F_2,\\
\dot{\xi} &= \mu F_3,\\
\dot{\varphi} &= 1 + \mu\xi,
\end{aligned}
\tag{4.3}
$$

where

$$
\begin{aligned}
F_1 &= \nu\cos\varphi - (h + h_1)x_1 - h_1 y,\\
F_2 &= \nu\sin\varphi - (h + h_1)y_1 + h_1 x,\\
F_3 &= \Delta - \gamma\xi - \lambda(x\sin\varphi - y\cos\varphi) - \chi(x_1 y - y_1 x) - \chi\left(x^2 + y^2\right).
\end{aligned}
$$

Terms of the order of μ^2 are omitted in system (4.3) as they do not affect the first approximation of the averaged system and, therefore, the dynamics of system (4.1).

System (4.3) represents a rotator loaded by a pair of oscillators. The ratio $(T - q\omega_0)\big/T_0 = \mu\Delta$ corresponds to the zone of main resonance: $(T - q\omega_0)\big/T_0 = \left(T\big/q - \omega_0\right)q\big/T_0 = (\Omega_0 - \omega_0)q\big/T_0 = \mu\Delta$, i.e. in the vicinity of frequency $\Omega_0 \approx \omega_0$, where Ω_0 is a partial frequency of rotation of the energy source.

Let us make a change of the variables in system (4.3):

$$x = u_1\sin\varphi + v_1\cos\varphi, \quad x_1 = u_1\cos\varphi - v_1\sin\varphi,$$
$$y = u_2\sin\varphi + v_2\cos\varphi, \quad y_1 = u_2\cos\varphi - v_2\sin\varphi.$$

As a result, obtained is the equivalent system in a standard form:

$$\dot{u}_1 = \mu(v_1\xi + F_1\cos\varphi),$$

$$\dot{v}_1 = -\mu(u_1\xi + F_1 \sin\varphi),$$
$$\dot{u}_2 = \mu(v_2\xi + F_2 \cos\varphi),$$
$$\dot{v}_2 = -\mu(u_2\xi + F_2 \sin\varphi),$$
$$\dot{\xi} = \mu F_3,$$
$$\dot{\varphi} = 1 + \mu\xi. \tag{4.4}$$

Having averaged system (4.4) over a fast-spinning phase, we obtain a system of the form

$$\dot{u}_1 = \mu\left(-\frac{h+h_1}{2}u_1 - \frac{h_1}{2}v_2 + v_1\xi + \frac{v}{2}\right),$$
$$\dot{v}_1 = \mu\left(-\frac{h+h_1}{2}v_1 + \frac{h_1}{2}u_2 - u_1\xi\right),$$
$$\dot{u}_2 = \mu\left(-\frac{h+h_1}{2}u_2 + \frac{h_1}{2}v_1 + v_2\xi\right),$$
$$\dot{v}_2 = \mu\left(-\frac{h+h_1}{2}v_2 - \frac{h_1}{2}u_1 - u_2\xi - \frac{v}{2}\right),$$
$$\dot{\xi} = \mu\left(\Delta - \gamma\xi - \frac{\lambda}{2}(u_1 - v_2) - \frac{\chi}{2}\left[(u_1 + v_2)^2 + (v_1 - u_2)^2\right]\right),$$
$$\dot{\varphi} = 1 + \mu\xi. \tag{4.5}$$

We have kept old notations for the averaged variables. One can see that a subsystem of the first five equations is independent of the sixth one and, therefore, can be studied separately. This subsystem has one important property that significantly simplifies our further study: in the phase space $G(u_1, u_2, v_1, v_2, \xi)$ of system (4.5), there exists a unique globally asymptotically stable invariant manifold $M = \{u_1 = -v_2, v_1 = u_2\}$.

To prove this statement, we introduce new variables $u_1 + v_2 = x_n, v_1 - u_2 = y_n$. (for such a change, the study of properties of the manifold is reduced to the study of properties of equilibria $u_1 + v_2 = x_n = 0, v_1 - u_2 = y_n = 0$. Further, subscript n is omitted.) to obtain the following system:

$$\dot{x} = -\frac{h+2h_1}{2}x + y\xi,$$
$$\dot{y} = -\frac{h+2h_1}{2}y - x\xi. \tag{4.6}$$

Evidently, equilibrium $x = 0, y = 0$ of system (4.6) is unique. On the other hand, a derivative of the Lyapunov function $V = x^2 + y^2$, taken along trajectories of the system $\dot{V} = -(h + 2h_1)(x^2 + y^2) \leq 0, \forall(u_1, u_2, v_1, v_2, \xi) \in G$, i.e. remains negative, which proves our statement.

Due to this property, all phase trajectories of system (4.5) are exponentially attracted by hyper-plane $M = \{u_1 = -v_2, v_1 = u_2\}$ and remain in its arbitrary, small vicinity for $\tau \to \infty$. This allows us to consider the system at the manifold instead of at system (4.5). The manifold is filled with trajectories of a three-dimensional dynamical system of the form

$$\dot{u} = \mu\left(-\frac{h}{2}u + v\xi + \frac{v}{2}\right),$$

$$\dot{v} = \mu\left(-\frac{h}{2}v - u\xi\right),$$

$$\dot{\xi} = \mu(\Delta - \gamma\xi - \lambda u). \tag{4.7}$$

The dynamics of system (4.5) is fully defined by the dynamics of system (4.7).

For the rotation characteristic of the rotator, we find that $\Omega = 1 + \mu\langle\xi^*(t, t_0)\rangle_t$, where $\xi^*(t, t_0)$ is the solution that corresponds to a trajectory of a limit set of the system realized for given initial conditions. We will show only a part of the rotation characteristic that corresponds to the resonance zone and, since $T \sim \Delta$ (linear proportion) and $\Omega \sim \langle\xi^*(t)\rangle_t$ (also a linear proportion), we will use dependency $\Delta = f(\langle\xi^*\rangle_t)$. This function contains all qualitative features of a real rotation characteristic. Both dependencies can be easily recalculated to each other.

Definition The function $A(\Omega) = \max_i \sqrt{x_*^2 + y_*^2}$ defined under parameter space and space of initial conditions, where $x_* = x_*(t, t_0)$, $y_* = y_*(t, t_0)$, is the limit solution for a system that corresponds to a given initial condition and is called the resonance characteristic of a flexible rotor.

Physically, the resonance characteristic defines a maximal displacement of the flexible rotor with respect to the unperturbed axis versus the frequency of rotation. In particular, if the motor (the energy source) has an unlimited power and its frequency is independent of the loading, then a resonance characteristic represents an amplitude-frequency characteristic of the oscillatory system.

Similarly to how it has been done for the rotation characteristic, we will present the resonance characteristic in the resonance domain only in the form $A = A(\Delta)$. Both characteristics are related to each other since $A = A(\Delta)$ is the resonance characteristic and $\Delta = \Delta(\Omega)$ is the rotation characteristic. The geometric form of the resonance characteristic and stability of its branches are determined by the form of the rotation characteristic and its stability.

Properties of the averaged system, rotation characteristic, and resonance characteristic. Let us perform a transformation of time of the form $\mu\tau = \tau_n$ and introduce more convenient notations: $\xi = x, u = y, v = z$. This reduces system (4.7) to an equivalent system of the form

$$\dot{x} = -\gamma x - \lambda y + \Delta,$$

$$\dot{y} = -\frac{h}{2}y + xz + \frac{v}{2},$$

$$\dot{z} = -\frac{h}{2}z - xy. \tag{4.8}$$

Let us list properties of system (4.8).

1. All solutions of the system are bounded. This is established with the help of quadratic form $V = x^2/2 + y^2/2 + (z - \lambda)^2/2$. Its derivative, taken along the vector field of the system $\dot{V} = -\gamma x^2 - \frac{h}{2}y^2 - \frac{h}{2}z^2 + \Delta x + \frac{v}{2}y + \frac{\lambda h}{2}z$, is negative outside some sphere R^2 (it is evident). This means that in the phase space $G(x, y, z)$ there exists a sphere $V \leq C^2$, that encloses sphere R^2. All phase trajectories get inside this sphere and remain there, i.e. $\|x\|, \|y\|, \|z\| < C < \infty$ for $t \to \infty$.

2. Depending on the parameters, system (4.8) has up to three equilibria, whose coordinates represent solutions of equation

$$f = \Delta = \gamma\omega + \frac{\lambda v h/4}{\omega^2 + h^2/4}. \tag{4.9}$$

Here $x_0 = \omega$, x_0 is a coordinate of the equilibrium.

Note that curve $f(\omega)$ represents a part of the rotation characteristic that corresponds to equilibria of system (4.8) [limit cycles of system (4.1)]. The resonance characteristic that corresponds to the equilibria has the form

$$A(\Delta) = \frac{v/2}{\sqrt{(\omega(\Delta))^2 + h^2/4}}. \tag{4.10}$$

It is not difficult to establish that for $\frac{\gamma h^2}{\lambda v} < \frac{3\sqrt{3}}{4}$ the rotation characteristic has two extrema (three equilibria for values of Δ contained between the extrema), while for $\frac{\gamma h^2}{\lambda v} > \frac{3\sqrt{3}}{4}$ the rotation characteristic is a one-one function (there is just one equilibrium for all Δ).

3. If the system has one equilibrium for all values of parameter Δ, then it is globally stable. This can be proven as follows.

Suppose that $O(x_0, y_0, z_0)$ is an equilibrium of the system under consideration. Using a change of the variables of the form $x = x_0 + u$, $y = y_0 + v$, $z = z_0 + w$, we obtain the following system

$$\dot{u} = -\gamma u - \lambda v,$$

$$\dot{v} = -\frac{h}{2}v + x_0 w + z_0 u + uw,$$

$$\dot{w} = -\frac{h}{2}w - x_0 v - y_0 u - uv. \tag{4.11}$$

The derivative of the Lyapunov function $V = \frac{1}{2}(mu^2 + v^2 + w^2)$ taken along trajectories of system (4.11) has the form $\dot{V} = -\frac{h}{2}(\alpha_1 u + v)^2 - \frac{h}{2}(\alpha_2 u + w)^2 \leq 0$, $\forall(u, v, w) \in G$, i.e. it is negative in the whole phase space; $\alpha_1 = (m\lambda - z_0)/h$,

$\alpha_2 = y_0/h$, m is a positive root of equation $\lambda^2 m^2 - 2(\lambda z_0 + h\gamma)m + y_0^2 + z_0^2 = 0$. The condition of existence of a positive root coincides with the condition of the rotation characteristic to be a one-one function. One-one rotation characteristics (see Fig. 4.2a) correspond to one-one resonance characteristics (see Fig. 4.3a).

4. Stability of equilibria in the case of not one-one rotation characteristics. A characteristic equation of the system in variations with respect to an arbitrary equilibrium $O(x_0, y_0, z_0)$ has the form:

$$p^3 + a_0 p^2 + a_1 p + a_2 = 0,$$

where $a_0 = h + \gamma$, $a_1 = h^2/4 + \omega^2 - \frac{\lambda v}{2}\frac{\omega}{h^2/4+\omega^2} + \gamma h$, $a_2 = \gamma\left(h^2/4 + \omega^2\right) - \frac{\lambda v h}{2}\frac{\omega}{h^2/4+\omega^2}$. The Gurvitz conditions of stability are expressed using the following

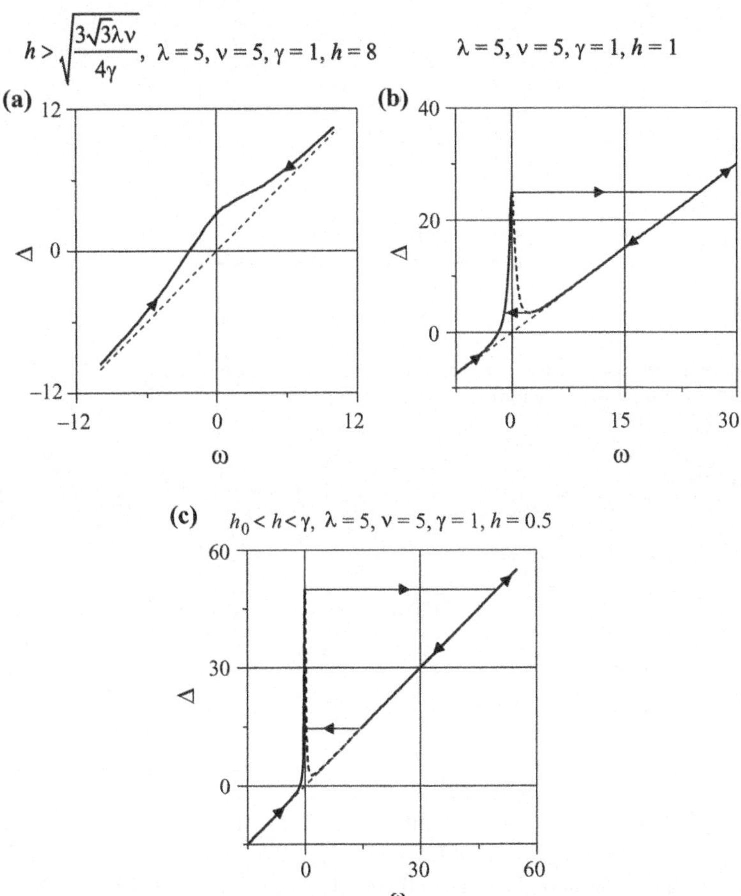

Fig. 4.2 Types of rotation characteristic in the resonance zone

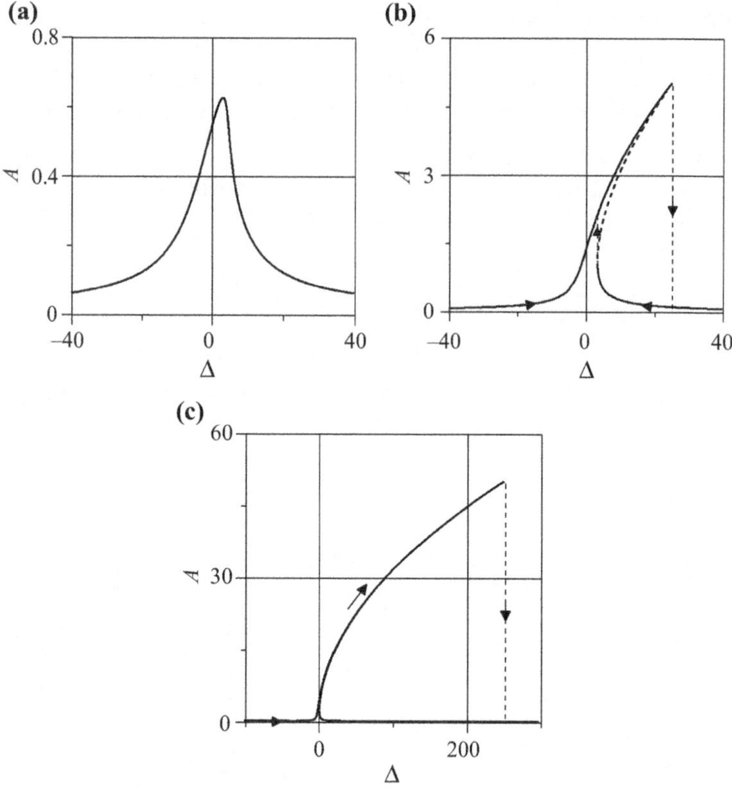

Fig. 4.3 Types of the resonance characteristic in the zone of resonance for $\lambda = 5$, $\nu = 5$, $\gamma = 1$, and **a** for $h = 8$; **b** for $h = 1$; **c** for $h = 0.1$

inequalities:

$$a_1 = f_1(\omega) = h^2\big/4 + \omega^2 + \gamma h - \frac{\lambda \nu}{2}\frac{\omega}{h^2\big/4 + \omega^2} > 0,$$

$$a_2 = f_2(\omega) = \gamma\big(h^2\big/4 + \omega^2\big) - \frac{\lambda \nu h}{2}\frac{\omega}{h^2\big/4 + \omega^2} = \big(h^2\big/4 + \omega^2\big)\Delta'_\omega > 0,$$

$$a_0 a_1 - a_2 = f_3(\omega) = h\big(h^2\big/4 + \omega^2\big) + \gamma h(h + \gamma) - \frac{\lambda \nu \gamma}{2}\frac{\omega}{h^2\big/4 + \omega^2} > 0.$$

$$(4.12)$$

It follows from the second inequality in (4.12) that the negatively sloped part of
the rotation characteristic is unstable for any $\Delta'_\omega < 0 \Rightarrow a_2 < 0$. Each point of this
part of the curve corresponds to a saddle equilibrium [7].

5. For $h \geq \gamma$, both positively sloped parts of the rotation characteristic are stable at all of their points. This can be established in the following way: since $\Delta'_\omega > 0$, then $a_2 > 0$ and the following inequality is satisfied:

$$\gamma\left(h^2/4 + \omega^2\right) > \frac{\lambda \nu h}{2} \frac{\omega}{h^2/4 + \omega^2}.$$

Using this inequality, we obtain:

$$a_1 = f_1(\omega) = h^2/4 + \omega^2 + \gamma h - \frac{\lambda \nu}{2} \frac{\omega}{h^2/4 + \omega^2} > h^2/4 + \omega^2$$

$$+ \gamma h - \frac{\gamma}{h}\left(h^2/4 + \omega^2\right) = \frac{h - \gamma}{h}\left(h^2/4 + \omega^2\right) + \gamma h > 0,$$

$$a_0 a_1 - a_2 = f_3(\omega) > h\left(h^2/4 + \omega^2\right) + \gamma h(h + \gamma) - \frac{\gamma^2}{h}\left(h^2/4 + \omega^2\right)$$

$$= \frac{h^2 - \gamma^2}{h}\left(h^2/4 + \omega^2\right) + \gamma h(h + \gamma) > 0.$$

Thus, for all points of positively sloped parts of the rotation characteristic, conditions of stability are satisfied for $h \geq \gamma$. In the case under consideration, equilibria that correspond to points of positively sloped parts are either stable focuses or stable nodes. At the extrema of the rotation characteristics, a bifurcation of merging of the equilibria, followed by the formation of a saddle-node and its further disappearance during a variation of parameter Δ, occurs.

6. Stability of positively sloped parts of the rotation characteristic for $h < \gamma$. Note that in the interval $\omega < 0$, the conditions of stability are satisfied for all points of the rotation characteristic. Also note that conditions of stability [first and third inequalities in (4.12)] improve for increased $h\left(\frac{\partial f_1}{\partial h} > 0, \frac{\partial f_3}{\partial h} > 0\right)$, which corresponds to the physical meaning of this parameter.

Suppose that $h \ll 1$. In this case, abscissae of the left (max.) and right (min.) extrema are defined via asymptotic formulae:

$$\omega_1 = \frac{h^3 \gamma}{8 \lambda \nu}, \qquad \omega_2 = \left(\frac{h \lambda \nu}{2\gamma}\right)^{1/3}.$$

Apart from that, for $\omega < \omega_1$, the following asymptotic formulae are valid:

$$f_1 = \gamma h - 2\lambda \nu \omega / h^2 > 0 \Rightarrow \omega < \gamma h^3/2\lambda \nu = 4\omega_1,$$
$$f_3 = \gamma^2 h - 2\lambda \nu \gamma \omega / h^2 > 0 \Rightarrow \omega < \gamma h^3/2\lambda \nu = 4\omega_1.$$

Thus, in the case under consideration, conditions of stability are satisfied for all points of the left-hand positively sloped part of the rotation characteristic. Due to the monotonous growth of functions f_1 and f_3 by parameter h, these conditions are valid for any h from the considered interval.

For the points of the right-hand positively sloped part of the curve $\omega > \omega_2$, we obtain asymptotic formulae of the form

$$f_1 = \omega^2 - \lambda v/2\omega > 0 \Rightarrow \omega > \left(\lambda v/2\right)^{1/3} = \omega_3,$$
$$f_3 = \gamma^2 h - \lambda v \gamma/2\omega > 0 \Rightarrow \omega > \lambda v/2\gamma h = \omega_4.$$

By varying parameter ω from higher values towards the lower ones, the equilibrium at the positively sloped part loses stability, becomes a saddle-focus, and remains as such for $\omega_3 < \omega < \omega_4$. The loss of stability happens due to the fact that an unstable limit cycle (which originates from the separatrix loop of saddle equilibrium) gets stuck in the equilibrium. Further, this equilibrium becomes a saddle [7] and remains as such for $\omega_2 < \omega < \omega_3$ until its disappearance in the minimum point of the rotation characteristic. Since f_1 and f_3 are monotonous functions, there exists value $h = h_0 < \gamma$ such that at the interval $h < h_0$, a part of the right-hand branch of the rotation characteristic is unstable.

Based on the studied properties of the averaged system, let us present a gallery of qualitatively different types of the rotation characteristic and the resonance characteristic. Let us also describe the dynamics of the system during the transition through resonance for different values of the damping coefficient. We assume that the torque is varied quasi-statically, while other parameters are kept constant.

(1) $h > \sqrt{\frac{3\sqrt{3}\lambda v}{4\gamma}}$. In this case, both the rotation and resonance characteristics are one-one functions. Examples of such characteristics are shown in Figs. 4.2a and 4.3a, respectively. For a high dissipation, the resonant properties of the shaft are weak and passage through the resonance in the forward, as well as in the backward, directions occurs along the same curve (i.e. there is no hysteresis) without any jumps in frequency of rotation.

(2) $h_0 \leq h < \sqrt{\frac{3\sqrt{3}\lambda v}{4\gamma}}$. Examples of rotation and resonance characteristics are shown in Figs. 4.2b and 4.3b, respectively. In this case, the rotation characteristic has two extrema. A negatively sloped part of the rotation characteristic is unstable, while both positively sloped parts are stable. During the increase of the torque (startup process, see arrows directed towards the right-hand side), a sticking of the frequency of rotation occurs as soon as a resonance frequency of the shaft is approached (the energy of the motor is transferred into the shaft vibrations). As soon as the torque reaches the value that corresponds to the left extremum of the rotation characteristic, a frequency jump occurs towards the right-hand branch that corresponds to a stable regime of rotations. For the opposite change of torque (shutdown process, see arrows directed towards the left-hand side), a frequency jump occurs from the left extremum of rotation characteristic towards the stable point of the right branch. This corresponds to motions along the resonance characteristic as shown in Fig. 4.3b by arrows.

(3) $h < h_0$. Examples of rotation and resonance characteristics are shown in Figs. 4.2c and 4.3c, respectively. The phenomenon of the sticking of the frequency of rotations becomes more pronounced. Besides that, during the shutdown process, a frequency jump towards the left-hand side occurs, not from the minimum of the rotation characteristic, but from point $\omega = \omega_4$ of the right-hand positively sloped branch. The resonance characteristic becomes very narrow so that one can see it in Fig. 4.3c where the positively sloped and negatively sloped branches almost coincide visually.

Let us estimate jumps of the frequency of rotation and of the amplitude of lateral vibrations for small values of h. A frequency jump to the right happens at the extremum of the rotation characteristic from the point $\omega \approx 0$. In this case $\Delta(0) = \lambda v/h$. The jump occurs to the point of the rotation characteristic that almost lies at the asymptote $\Delta = \gamma\omega$, to the point $\omega = \frac{\lambda v}{\gamma h}$, i.e. $\delta\omega = \frac{\lambda v}{\gamma h}$. Using the formulae relating dimensional and dimensionless variables and parameters, we obtain $\delta\Omega = \frac{c^2 e^2}{\varepsilon q \omega_0}$. For small values of parameter ε, the amplitude of the jump is significant.

It is evident that $\max A(\Delta) = A(\Delta_0) = \frac{v}{h}$ for $\Delta_0 = \frac{\lambda v}{h}$. Expressed in physical parameters: $\max A\left(T/q\right) = A(\Omega_0) = \frac{ec}{\varepsilon\omega_0}$, and $\Omega_0 = \omega_0 + \frac{e^2 c^2}{q\varepsilon\omega_0}$.

4.2 Damping Lateral Vibrations in Rotary Machinery Using Motor Speed Modulation

It is known that amplitudes of vibrations of rotating shafts under resonance conditions can reach large and even unacceptably high values. As noted earlier, this can happen when the shaft is accelerated to the operating speed and passes resonance zones, such as at the operating frequency, if it is close to one of the resonant frequencies. The effect of increasing the amplitude of vibrations at the operational frequency can also manifest as a result of changes in the shaft parameters during operation. For instance, in the oil and gas industry this can occur during long-term operation of electric submersible pumps (ESP), when scale deposits on the impeller blades. This increases the total mass of the shaft and shifts down the spectrum of the resonant frequencies, as well as causes an increase of the imbalance. An increase in the amplitude of bending vibrations can be facilitated by wear of the (sliding) bearings as a result of changes in parameters of the lubricating layer and changes in their geometry during long-term service. The wear and tear of the equipment can lead to unacceptable vibrations, which ultimately may result in catastrophic equipment failure.

Application of mechanical vibration dampers might be ineffective due to two reasons: (1) the system may be inaccessible during exploitation, which prevents vibration control and replacement of failed components; (2) no possibility to change the frequency band of vibration dampers. In what follows, we will study a possibility to damp lateral vibrations of rotary systems by controlling the frequency of rotations [8, 9], i.e. controlling the speed of the asynchronous motor.

Consider a dynamical system that governs lateral vibrations of a rotor coupled to the energy source (motor) of unlimited power. When the rotor passes through a resonance zone, a harmonic component is added to a constant torque. Such a harmonic additive may also be added in the normal operating regime in a case when, for instance, the control system shows an undesirable increase in vibration. In these cases, the constant frequency of rotation experiences harmonic modulation and the governing equations have the form:

$$\ddot{x} + \frac{\varepsilon\omega_0}{c}\dot{x} + x + \frac{k\omega_0}{c}(\dot{x} + \dot{\varphi}y) = \frac{e}{A_0}\cos\varphi,$$

$$\ddot{y} + \frac{\varepsilon\omega_0}{c}\dot{y} + y + \frac{k\omega_0}{c}(\dot{y} - \dot{\varphi}x) = \frac{e}{A_0}\sin\varphi,$$

$$\dot{\varphi} = \Omega + An\Omega\cos(n\Omega\tau + \psi_0). \tag{4.13}$$

Variables, parameters, and time in system (4.13) are dimensionless; $\Omega = \omega/\omega_0$. The differentiation is done over dimensionless time $\omega_0 t = \tau$.

We suppose that coefficients of external and internal damping as well as rotor eccentricity are small: $c/m = \omega_0^2$, $\varepsilon\omega_0/c = \mu h$, $k\omega_0/c = \mu h_1$, $e/A_0 = \mu\nu$, where $\mu \ll 1$ is a certain small parameter. System (4.13) takes the form

$$\dot{x} = x_1,$$

$$\dot{x}_1 = -\Omega^2 x + \mu F_1,$$

$$\dot{y} = y_1,$$

$$\dot{y}_1 = -\Omega^2 y + \mu F_2,$$

$$\dot{\psi} = \Omega. \tag{4.14}$$

Here $F_1 = \nu\cos\varphi - (h + h_1)x_1 - h_1\dot{\varphi}y + \Delta x$, $F_2 = \nu\sin\varphi - (h + h_1)y_1 + h_1\dot{\varphi}x + \Delta y$, $\varphi = \psi + A\sin(n\psi + \psi_0)$.

Our main goal is to answer the following question: would it be possible to choose the parameters of modulation A, n, ψ_0, in such way that the amplitude of lateral vibrations would be minimized and, if so, then how should these parameters be chosen?

The physical view of the problem can be as follows. System (4.13) and system (4.14), respectively, are linear non-autonomous systems of two oscillators. Each of these oscillators represents a filter of the resonant frequency Ω, which represents first harmonics in the spectrum of the external force:

$$\cos\varphi = \cos(\psi + A\sin(n\psi + \psi_0)), \ \sin\varphi = \sin(\psi + A\sin(n\psi + \psi_0)),$$

$$\psi = \Omega\tau.$$

These functions may be decomposed into the Fourier series with coefficients $J_k(A)$ representing Bessel functions of the first kind of integer argument. Naturally, since the model under consideration is linear, its solutions can also be represented as such series. If the first harmonic that has the highest amplitude would be damped by a proper choice of the parameters of modulation, then the amplitudes of the remaining harmonics would have values on the order of the external force, i.e. $\sim\mu \ll 1$. It means that the maximum radial displacements of the shaft $x_{\max}, y_{\max}$ would have the same order: $x_{\max}, y_{\max} \sim \mu$ (note that $x(\tau), y(\tau)$ are the multi-frequency functions). Thus, our objective is to damp the first harmonic.

A direct solution of this problem consists of searching for the solution of system (4.14) in a form of the Fourier series with undetermined coefficients, determination of these coefficients, and minimization of the amplitude of the first harmonic. At the same time, stability of the solution must be ensured. In general, such approach is quite difficult to carry out. For this reason, let us use the method of the averaging.

$$x = u_1 \sin\psi + v_1 \cos\psi, \quad x_1 = (u_1 \cos\psi - v_1 \sin\psi)\Omega,$$
$$y = u_2 \sin\psi + v_2 \cos\psi, \quad y_1 = (u_2 \cos\psi - v_2 \sin\psi)\Omega$$

We reduce system (4.14) to the standard form

$$\dot{u}_1 = \mu F_1 \cos\psi,$$
$$\dot{v}_1 = -\mu F_1 \sin\psi,$$
$$\dot{u}_2 = \mu F_2 \cos\psi,$$
$$\dot{v}_2 = -\mu F_2 \sin\psi,$$
$$\dot{\psi} = \Omega. \tag{4.15}$$

Prior to the procedure of the averaging, let us note that mean values of the variables

$$\langle \dot{\varphi} y \sin\psi \rangle_\psi, \ \langle \dot{\varphi} y \cos\psi \rangle_\psi, \ \langle \dot{\varphi} x \sin\psi \rangle_\psi, \ \langle \dot{\varphi} x \sin\psi \rangle_\psi,$$
$$\langle \cos(\psi + A \sin(n\psi + \psi_0)) \sin\psi \rangle_\psi, \ \langle \cos(\psi + A \sin(n\psi + \psi_0)) \cos\psi \rangle_\psi,$$
$$\langle \sin\psi(\psi + A \sin(n\psi + \psi_0)) \sin\psi \rangle_\psi, \ \langle \sin(\psi + A \sin(n\psi + \psi_0)) \cos\psi \rangle_\psi$$

significantly depend on parameter n. For this reason, we consider several cases.

1. *Any integer value of parameter $n > 2$, as well as any irrational number.* By averaging system (4.15) over a fast-spinning phase ψ and by transforming time: $\mu\tau = \tau_n$, we obtain equations in the first approximation with respect to the small parameter of the form

$$\dot{u}_1 = -\frac{h + h_1}{2}u_1 - \frac{h_1}{2}v_2 + \frac{\Delta}{2}v_1 + \frac{1}{2}\nu J_0(A),$$
$$\dot{v}_1 = -\frac{h + h_1}{2}v_1 + \frac{h_1}{2}u_2 - \frac{\Delta}{2}u_1,$$

$$\dot{u}_2 = -\frac{h+h_1}{2}u_2 + \frac{h_1}{2}v_1 + \frac{\Delta}{2}v_2,$$

$$\dot{v}_2 = -\frac{h+h_1}{2}v_2 - \frac{h_1}{2}u_1 - \frac{\Delta}{2}u_2 - \frac{1}{2}vJ_0(A). \tag{4.16}$$

Here, $J_0(A)$ is the Bessel function of the first kind. In this case, the equations are independent of phase shift ψ_0.

Consider value $A^* = \sqrt{u_{10}^2 + u_{20}^2 + v_{10}^2 + v_{20}^2}$, where u_{10}, u_{20}, v_{10}, v_{20} are the coordinates of the equilibrium of linear system (4.16), which represents the amplitude of the first harmonic. Our goal is to minimize this value.

Similarly to Sect. 4.1, we note that system (4.16) has an invariant manifold $M = \{u_1 = -v_2, u_2 = v_1\}$, which is stable. This can be proven in the same way as before: consider the system at the manifold $M = \{u_1 = -v_2, u_2 = v_1 = v\}$

$$\dot{u} = -\frac{h}{2}u + \frac{\Delta}{2}v + \frac{v}{2}J_0(A),$$

$$\dot{v} = -\frac{h}{2}v - \frac{\Delta}{2}u. \tag{4.17}$$

Values of coordinates of equilibrium (u_0, v_0) of system (4.17) are proportional to $J_0(A) = 0$, and, therefore, the amplitude of the first harmonic is minimal for the minimal values of $|J_0(A)|$. In particular, it equals zero (which means full damping) for all values of A, for which $J_0(A) = 0$. It is well-known that Bessel function $J_0(A)$ has an infinite number of zeros (see Fig. 1.8). For instance, the first zero corresponds to $A = 2.4$, which corresponds to the minimum value of A. Substituting $A = 2.4$, into system (4.17), we obtain equilibrium $u = 0$, $v = 0$, which is, obviously, stable.

Therefore, by setting parameters of the frequency modulation: $A = 2.4$, any integer $n > 2$ or any irrational one, as well as any value of the phase shift ψ_0 (for instance, $\psi_0 = 0$), we obtain a full damping of the first harmonic of bending vibrations of the rotor. In this case, the amplitude of the bending vibrations $\sim\mu \ll 1$.

Numerical study. Since internal damping does not affect the amplitude of lateral vibrations, instead of system (4.13) we have studied an equivalent system of the form

$$m\ddot{x} + \varepsilon\dot{x} + cx = ce\cos\varphi,$$

$$m\ddot{y} + \varepsilon\dot{y} + cy = ce\sin\varphi,$$

$$\dot{\varphi} = \Omega + An\Omega\cos(n\Omega t).$$

The following parameters have been chosen: $m = 1$, $\varepsilon = 0.1$, $c = 25$, $e = 1$ for different parameters of the damping system, i.e. parameters of the modulation. The following two cases are of interest: (1) performance of the damping system when the motor speed is fixed and resonates with the frequency of the lateral vibrations of the shaft; (2) effectiveness of the damping of transverse vibrations of the shaft in the case of startup and acceleration of the engine followed by a passage of its speed through the resonant zone, while the damping system is constantly turned on.

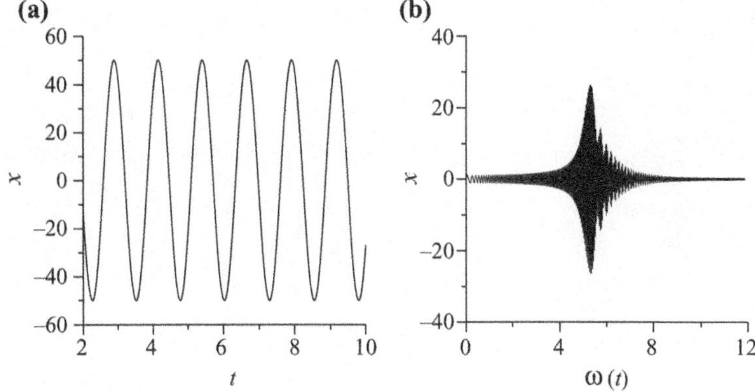

Fig. 4.4 Time-history of solution $x = x^*(t)$ of system (4.13) in case, when the damping system is inactive (**a**); lateral displacement of the shaft $x(\omega(t))$ during the startup process and passage through the resonance zone (**b**)

Figure 4.4a shows a time history of solution $x = x^*(t)$ at the resonant frequency for the undamped case ($n = 0$). In this case, the frequency of rotation of the motor is constant and is equal to the resonant frequency of the shaft. Figure 4.4b shows lateral displacement of the shaft $x(\omega(t))$ during the startup process and bypass through the resonance zone when the damping system is turned off.

In what follows, we will estimate the effectiveness of the damping system by means of the damping ratio: the ratio between amplitudes of lateral vibrations of the shaft when the damping system is inactive (see Fig. 4.4) and those when the damping system is active. Also, we consider two cases: the first one, when the damping system turns on for a fixed (resonant) frequency of rotation; and the second one, during the startup and acceleration of the rotor followed by the passage through the resonance zone, when the damping system is continuously active.

Figure 4.5a shows a transient process of lateral vibrations of the shaft after the damping system has been activated using the following parameters: $n = 3$, $A = 2.4$ (a constant component of the frequency of rotations is fixed and is resonant). Figure 4.5b shows steady-state vibrations of the shaft as a result of the damping. Figure 4.5c shows lateral displacement of the shaft $x(\omega(t))$ during the startup of the rotor through the zone of resonance for the case when the damping system is constantly turned on. Calculations show that for a fixed frequency of rotation, the damping ratio equals 217.4. For the continuously active damping system, when the system passes through the resonance zone, the coefficient of damping equals 7.7. A significant difference between these coefficients is caused by a high Q-factor of the oscillatory system and, consequently, by a long-lasting process of attenuation of natural oscillations: the damping system does not fully perform during a relatively short time of passage through the resonance zone.

(2) *Modulation at the frequency of rotation of the rotor, $n = 1$*, In this case, the averaged system has the form

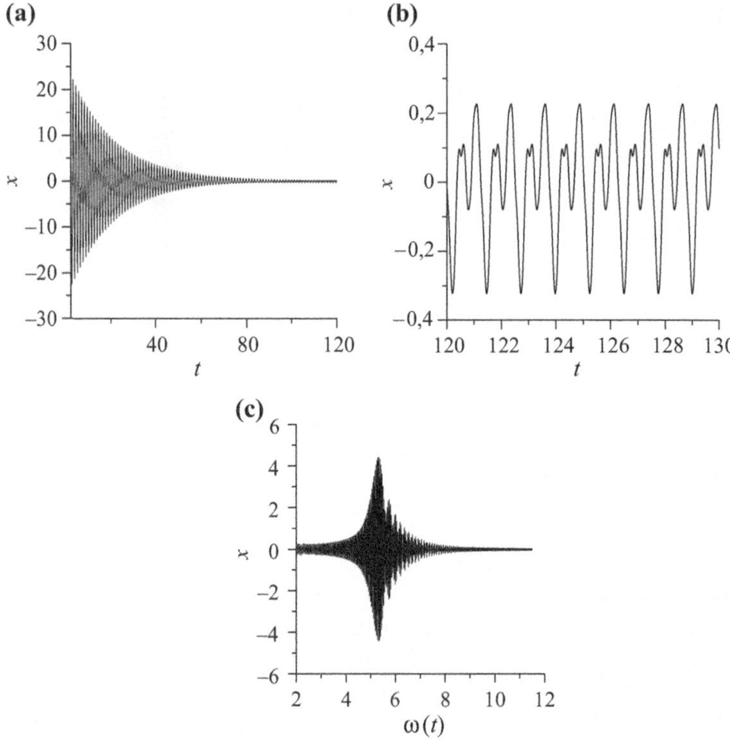

Fig. 4.5 A transient process of damping of vibrations (**a**); solution $x = x^*(t)$ as a result of the damping (**b**); lateral displacement $x(\omega(t))$ during the startup process and passage through the resonance zone during the constantly active damping system (**c**); $n = 3$, $A = 2.4$

$$\dot{u}_1 = -\frac{h + h_1}{2}u_1 - \frac{h_1}{2}v_2 + \frac{\Delta}{2}v_1 + \frac{\nu}{2}(J_0 + J_2 \cos 2\psi_0),$$

$$\dot{v}_1 = -\frac{h + h_1}{2}v_1 + \frac{h_1}{2}u_2 - \frac{\Delta}{2}u_1 + \frac{\nu}{2}J_2 \sin 2\psi_0,$$

$$\dot{u}_2 = -\frac{h + h_1}{2}u_2 + \frac{h_1}{2}v_1 + \frac{\Delta}{2}v_2 - \frac{\nu}{2}J_2 \sin 2\psi_0,$$

$$\dot{v}_2 = -\frac{h + h_1}{2}v_2 - \frac{h_1}{2}u_1 - \frac{\Delta}{2}u_2 - \frac{\nu}{2}(J_0 - J_2 \cos 2\psi_0). \tag{4.18}$$

Equilibrium of system (4.18) is stable (the corresponding homogeneous system has a stable integral manifold with stable system, defined at this manifold, see above). This equilibrium has zero coordinates independent of phase ψ_0, if $J_0 = J_2 = 0$ (normally, the initial phase is difficult to control). As Fig. 1.8 shows, the zeros of these Bessel functions (starting with the second one, $A \approx 5.1$) are located quite close to each other and, therefore, similar to the previous case, by a proper selection of

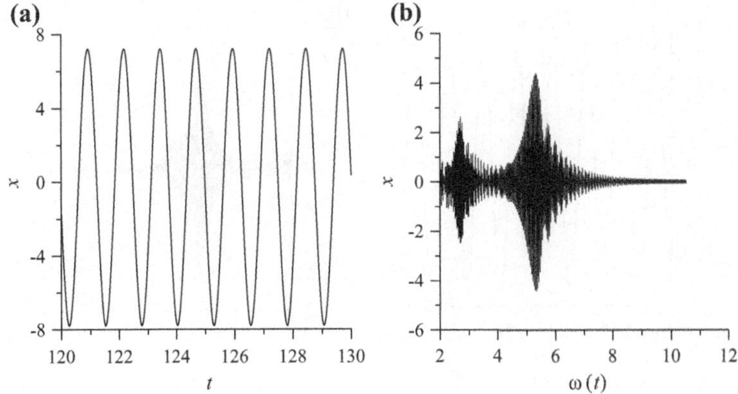

Fig. 4.6 Solution $x = x^*(t)$ as a result of the damping (**a**); lateral displacement of the shaft $x(\omega(t))$ during the startup process and passage through the resonance zone during the constantly active damping system (**b**); $n = 1$, $A = 5.1$

amplitude of modulation, one could obtain almost a complete damping of the first harmonic of lateral vibrations.

Numerical study. Figure 4.6 shows damped steady-state vibrations (Fig. 4.6a) and lateral displacement of the shaft $x(\omega(t))$ during the startup of the rotor and passage through the resonance zone with constantly active damping system (Fig. 4.6b) for $n = 1$, $A = 5.1$. In the first case, the damping ratio equals 33.3, while in the second one it equals 7.

(3) *Modulation at doubled frequency of rotation, $n = 2$.* In this case, the averaged system has the form

$$
\dot{u}_1 = -\frac{h + h_1}{2}u_1 - \frac{h_1}{2}v_2 + \frac{\Delta}{2}v_1 - \frac{h_1 A}{2}(-u_2 \sin \psi_0 + v_2 \cos \psi_0)
$$
$$
+ \frac{v}{2}(J_0 - J_1 \cos \psi_0),
$$
$$
\dot{v}_1 = -\frac{h + h_1}{2}v_1 + \frac{h_1}{2}u_2 - \frac{\Delta}{2}u_1 + \frac{h_1 A}{2}(-u_2 \cos \psi_0 - v_2 \sin \psi_0)
$$
$$
- \frac{v}{2}J_1 \sin \psi_0,
$$
$$
\dot{u}_2 = -\frac{h + h_1}{2}u_2 + \frac{h_1}{2}v_1 + \frac{\Delta}{2}v_2 + \frac{h_1 A}{2}(-u_1 \sin \psi_0 + v_1 \cos \psi_0)
$$
$$
+ \frac{v}{2}J_1 \sin \psi_0,
$$
$$
\dot{v}_2 = -\frac{h + h_1}{2}v_2 - \frac{h_1}{2}u_1 - \frac{\Delta}{2}u_2 - \frac{h_1 A}{2}(-u_1 \cos \psi_0 - v_1 \sin \psi_0)
$$
$$
- \frac{v}{2}(J_0 + J_1 \cos \psi_0). \tag{4.19}
$$

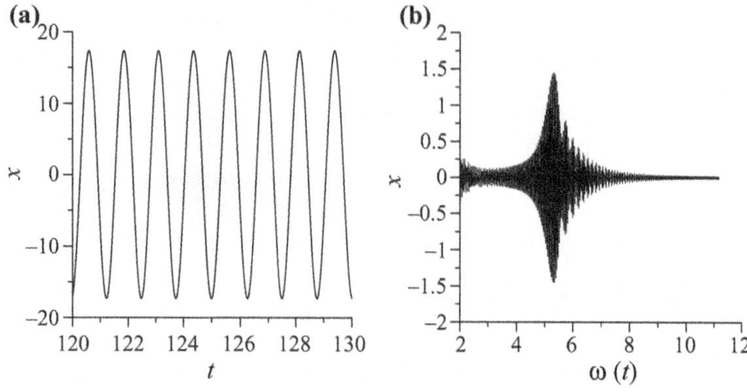

Fig. 4.7 Result of the damping of lateral vibrations in the case of stationary rotation of the rotor (**a**); lateral displacement of the shaft $x(\omega(t))$ during the startup process and passage through the resonance zone during the constantly active damping system (**b**); $n = 2$, $A = 5.5$

Using the same considerations as above, we conclude that for a full damping of the first harmonic, it is necessary that the following equality is fulfilled: $J_0 = J_1 = 0$. However, this equation does not have a solution including the approximate one that would have satisfactory precision (see Fig. 1.8). For this reason, it would be natural to select the value of the amplitude of the modulation, for which these functions have minimum values.

Figure 4.7 shows damped steady-state vibrations (see Fig. 4.7a) and lateral displacement of the shaft $x(\omega(t))$ during the startup of the rotor and passage through the resonance zone with constantly active damping system (Fig. 4.7b) for $n = 2$, $A = 5.5$.

Calculations show that if the damping system is turned on at the resonant frequency, then the damping ratio equals 2.85; while for a startup of the rotor and passage through the resonance zone for a constantly active damping system, the damping ratio equals 33.3.

As one can see, the principle of the damping of lateral vibrations of shafts by means of the modulation of its frequency of rotations works well. Of course, this method could be implemented as a fully automatic one.

4.3 Chaotic Torsional Vibrations of the Shaft in a System with Limited Power Source

In this section, we consider dynamics of a torsionally-elastic shaft driven by an energy source of limited power [10]. An asynchronous motor is taken as an example, but in general an engine of any other type could be chosen. We consider a simple model, when an elastic shaft, (distributed system) including other systems installed on it, is

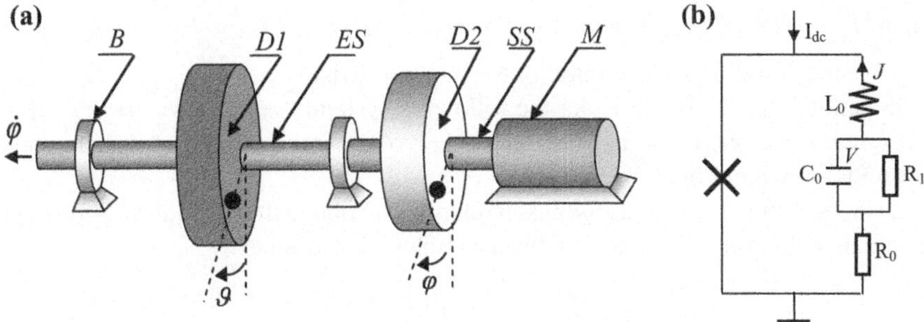

Fig. 4.8 Asynchronous motor with unbalanced disk and adduction disk of the elastic shaft (**a**): *M* is the motor, *ES* is the elastic shaft, *D1* is the unbalanced disk, *SS* is the solid shaft, *D2* is the adduction disk of elastic shaft, *B* is the bearing; an equivalent scheme of the system "*superconductive junction – resonator*" (**b**)

modeled as one disk with known moment of inertia. The disk is placed in the center of the shaft (one mode model). We will assume that all resistance forces are linear with respect to angular speed and elastic restoring force is linear with respect to the angle of twist. A scheme of this model is shown in Fig. 4.8a.

The Lagrange function for this system has the form

$$L = T + V = \frac{I_1}{2}\dot{\varphi}^2 + \frac{I_0}{2}\dot{\vartheta}^2 + mge(1 - \cos\varphi) + \frac{k}{2}(\varphi - \vartheta)^2,$$

where φ, ϑ are the angles of rotation of the imbalanced disk and adduction disk, respectively; I_1, I_0 are their moments of inertia, respectively; m is the imbalance, e is the eccentricity, and k is the torsional stiffness of the elastic shaft.

By assumption, the motor driving torque is a linear function of the rotational speed: $M_d = M_0 - \lambda\dot{\varphi}$. We also account for the fact that the whole system can operate inside a viscous environment (like electric submersible pumps). We assume that all moments of dissipative forces acting on the imbalanced disk are described by the term $-\lambda\dot{\varphi}$ in the expression for the torque. The moment of external forces acting on the adduction disk has the form $-\delta_0\dot{\vartheta}$. Also, we assume that there exists a hypothetical moment of forces of internal torsional friction $\varepsilon_0(\dot{\varphi} - \dot{\theta})$. Under these assumptions, the governing equations have the form

$$I_1\ddot{\varphi} + mge\sin\varphi + k(\varphi - \vartheta) = M_0 - \lambda\dot{\varphi},$$
$$I_0\ddot{\vartheta} + k(\vartheta - \varphi) = -\delta_0\dot{\vartheta} + \varepsilon_0(\dot{\varphi} - \dot{\vartheta}).$$

Introducing a new variable $\frac{k}{mge}(\vartheta - \varphi) = J$ and time $\frac{mge}{\lambda}\tau = \tau_n$, we reduce the system above to the form:

$$I\ddot{\varphi} + \dot{\varphi} + \sin\varphi = \gamma + J,$$
$$\ddot{J} + \delta\dot{J} + \omega_0^2 J = d\dot{\varphi} + b\ddot{\varphi}, \tag{4.20}$$

where $I = \frac{I_1 mge}{\lambda^2}$, $\gamma = \frac{M_0}{mge}$, $\delta = (\delta_0 + \varepsilon_0)I_0^{-1}\frac{\lambda}{mge}$, $\omega_0^2 = kI_0^{-1}\left(\frac{\lambda}{mge}\right)^2$, $b = -\frac{k}{mge}$, $d = -\frac{\delta_0}{\lambda}\omega_0^2$. Variable J has a sense of the angle of twist.

In parallel, let us also consider the following system: "*superconductive junction – resonator*" in one-mode approximation [11, 12]. An equivalent system for this system is shown in Fig. 4.8b.

Using Kirchhoff laws and Josephson relation, we obtain dimensionless governing equations with dimensionless variables, parameters, and time:

$$c\ddot{\varphi} + \dot{\varphi} + \sin\varphi = \gamma + J,$$
$$l_0\dot{J} + r_0 J - V = -\dot{\varphi},\ J = -c_0\dot{V} - r_1^{-1}V.$$

By excluding variable V, we obtain system of the form (4.20) with the following parameters:

$$\delta = \frac{c_0 r_0 r_1 + l_0}{c_0 l_0 r_1},\ \omega_0^2 = \frac{1}{c_0 l_0}\left(1 + \frac{r_0}{r_1}\right),\ b = -\frac{1}{l_0},\ d = -\frac{1}{c_0 l_0 r_1}.$$

Thus, we have a quantum-mechanical analogue for a mechanical system and vice versa.

Consider the dynamics of system (4.20) with the following conditions for parameters:

$$I^{-1} = \mu \ll 1,\ \delta = \mu h,\ \gamma + \frac{d}{\omega_0} - \omega_0 = \mu\Delta.$$

For these conditions, our system represents a quail-linear rotator coupled to a high-quality oscillator. The last equation determines the resonance zone. System (4.20) looks similar to system (3.1). However, we will show that system (4.20) has some qualitatively different dynamical properties.

To study this system, we use the same technique as has been used in the previous sections, except the form of changes of the variables will be different. For this reason, we will skip certain details that have already been outlined before. A system, which is equivalent to system (4.20), has the form

$$I\ddot{\varphi} + \dot{\varphi} + \sin\varphi = \gamma + J,$$
$$\dot{J} = W + b\dot{\varphi} - \mu h J,$$
$$\dot{W} = -\omega_0^2 J + d\dot{\varphi}$$

By applying changes of the variables

$$J = \theta\sin\varphi + \eta\cos\varphi + \frac{d}{\omega_0},$$
$$W = (\theta\cos\varphi - \eta\sin\varphi)\omega_0 - b\omega_0,$$

$$\dot{\varphi} = \omega_0 + \mu\, \Phi(\theta, \eta, \varphi, \xi) \tag{4.21}$$

We obtain an equivalent system in a standard form (3.4) with functions

$$\Theta = \left(\eta + b\sin\varphi + \frac{d}{\omega_0}\cos\varphi\right)\Phi - h(\theta\sin\varphi + \eta\cos\varphi)\sin\varphi + \frac{d}{\omega_0}\sin\varphi,$$

$$T = \left(b\cos\varphi - \theta - \frac{d}{\omega_0}\sin\varphi\right)\Phi - h(\theta\sin\varphi + \eta\cos\varphi)\cos\varphi + \frac{d}{\omega_0}\cos\varphi,$$

$$\Xi = \Delta - \left(\frac{\partial\Phi}{\partial\theta}\Theta + \frac{\partial\Phi}{\partial\eta}T + \frac{\partial\Phi}{\partial\varphi}\Phi + \Phi\right), \quad \Phi = \frac{1-\theta}{\omega_0}\cos\varphi + \frac{\eta}{\omega_0}\sin\varphi + \xi.$$

The averaged system has the same form as system (3.10):

$$\dot{\xi} = \mu(-b_1\xi + b_2\theta + b_3\eta + \Delta),$$
$$\dot{\theta} = \mu(-b_4\theta + b_5\eta + \eta\xi + b_6),$$
$$\dot{\eta} = \mu(-b_4\eta - b_5\theta - \theta\xi + b_7),$$
$$\dot{\varphi} = \omega_0 + \mu\xi$$

With the following parameters:

$$b_1 = 1 - \frac{d}{\omega_0^2},\, b_2 = 0,\, b_3 = \frac{1}{2\omega_0^2},\, b_4 = \frac{1}{2}\left(h + \frac{d}{\omega_0^2}\right),\, b_5 = \frac{b}{2\omega_0},$$

$$b_6 = \frac{d}{2\omega_0^2},\, b_7 = \frac{b}{2\omega_0}.$$

Further, a change of the variables of the form

$$x = b_4^{-1}(\xi + b_5), \quad y = \frac{b_3\eta + b_2\theta}{b_1 b_4} - \Lambda,$$

$$z = \frac{b_3\theta - b_2\eta}{b_1 b_4} + R, \quad \mu b_4 \tau = \tau_n \tag{4.22}$$

reduces the averaged system to a Lorenz type system of the form

$$\dot{x} = -\sigma(x - y) + \rho,$$
$$\dot{y} = -y + Rx - xz,$$
$$\dot{z} = -z + xy + \Lambda x, \tag{4.23}$$

whose parameters are expressed as follows:

$$\sigma = b_1/b_4 = \frac{2(\lambda + \delta_0)}{h\lambda - \delta_0}, \quad R = \frac{b_2 b_7 - b_3 b_6}{b_1 b_4^2} = \frac{\delta_0\lambda}{\omega_0^2(\lambda + \delta_0)(h\lambda - \delta_0)^2},$$

$$\Lambda = \frac{b_3 b_7 + b_2 b_6}{b_1 b_4^2} = -\frac{k\lambda^3}{mgl\omega_0^3(\lambda + \delta_0)(\lambda h - \delta_0)^2}$$

$$\rho = \left(\Delta + (b_1 b_4 b_5 + b_3 b_7 + b_2 b_6)b_4^{-1}\right)b_4^{-2}.$$

This system has the same properties as outlined in Sect. 3.2. The only difference is that in this case $R > 0$, and $\Lambda < 0$.

Proceeding from the fact that system (4.23) is the averaged system for system (4.20), we obtain statements regarding the conditions of existence of regular and chaotic torsional vibrations of the shaft. In particular, we state that there exist chaotic Feigenbaum and Lorentz attractors that are observed in the space of the Poincaré point mapping.

Numerical study. Figures 4.9 and 4.10 show galleries of phase portraits of system (4.20) at the involute of the phase cylinder as well as portraits of the point mapping. The point mapping of the form $(x, \theta, \eta)_{\varphi=\varphi_0} \rightarrow (\bar{x}, \bar{\theta}, \bar{\eta})_{\varphi=\varphi_0+2\pi}$ has been performed for the system equivalent to system (4.20) of the form

$$\dot{\varphi} = \omega_0 + \mu x,$$

$$\dot{x} = -\mu x - \sin\varphi + \theta\sin\varphi + \eta\cos\varphi + \gamma - \omega_0,$$

$$\dot{\theta} = \mu x\eta - \delta(\theta\sin\varphi + \eta\cos\varphi)\sin\varphi + \left(b\sin\varphi + \frac{d}{\omega_0}\cos\varphi\right)(\omega_0 + \mu x),$$

$$\dot{\eta} = -\mu x\theta - \delta(\theta\sin\varphi + \eta\cos\varphi)\cos\varphi + \left(b\cos\varphi - \frac{d}{\omega_0}\sin\varphi\right)(\omega_0 + \mu x).$$

Experiment 1 The following parameter set has been used: $d = -0.27508$, $\mu = 0.01$, $\delta = 0.0134$, $b = -0.005016$, $\omega_0 = 0.5$. Figure 4.9 shows a gallery of phase portraits and Poincaré mappings for the different values of parameter γ. Parameter $b = -0.005016$ is chosen to be small in order to correspond to a small value of parameter Λ of the averaged system. During the variation of parameter γ, the resonance zone is passed from left to right.

For $\gamma = 1.0448906$ and lower values, a periodic motion (limit cycle) is observed in system (4.20) (see Fig. 4.9a). Figure 4.9b shows a transient process of formation of a fixed point of the Poincaré mapping that represents a focus.

For the increased values of γ, the fixed point loses stability via the opposite Andronov-Hopf bifurcation, i.e. due to the sticking of the limit cycle in the equilibrium (in the phase space of the averaged system and space of the point mapping). In a four-dimensional phase space of system (4.20), a sticking of an unstable torus into a limit cycle occurs that is followed by instability "transfer" from the torus to the limit cycle. As a result of this bifurcation, our system suddenly jumps from the regime of periodic motions to the regime of chaotic vibrations, whose image is an asymmetric Lorenz attractor shown in Fig. 4.9c, d. In this case, both the instant frequency of the motor and the amplitude of torsional vibrations are changed chaotically. The situation is such that the affix that is staying most of the time in the vicinity of pre-resonance

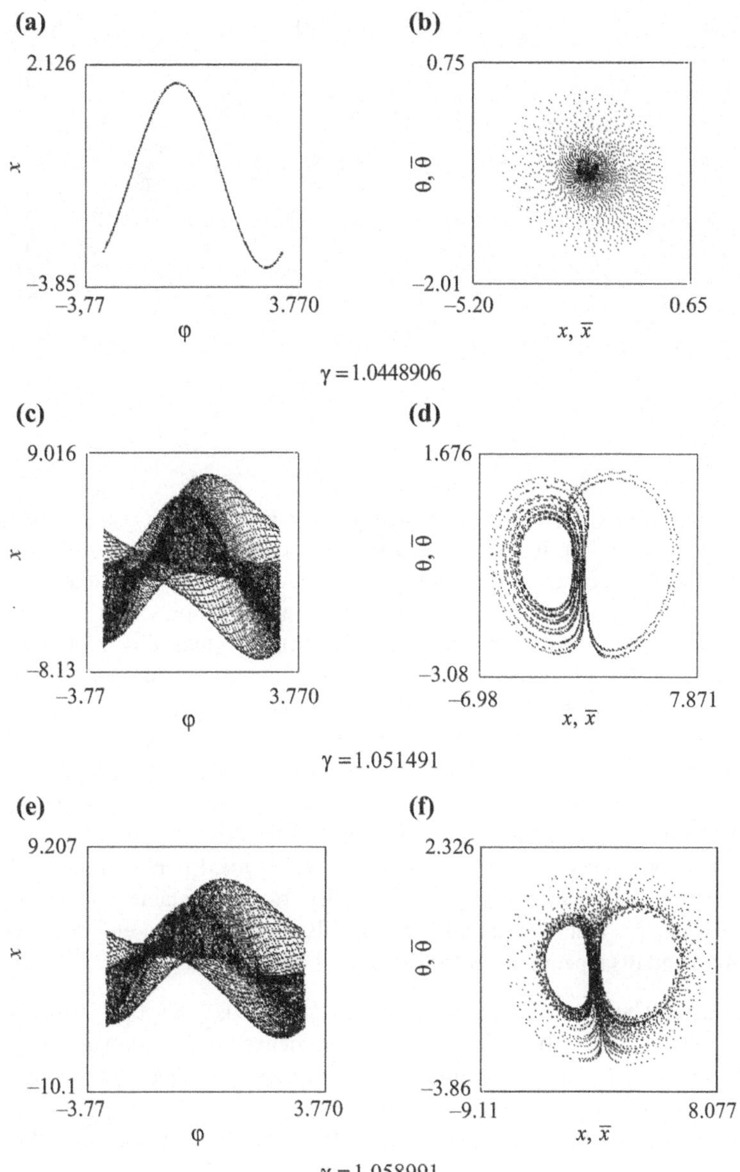

Fig. 4.9 The gallery of the phase portraits and portraits of the Poincaré mapping for Experiment 1 (continues on the next page)

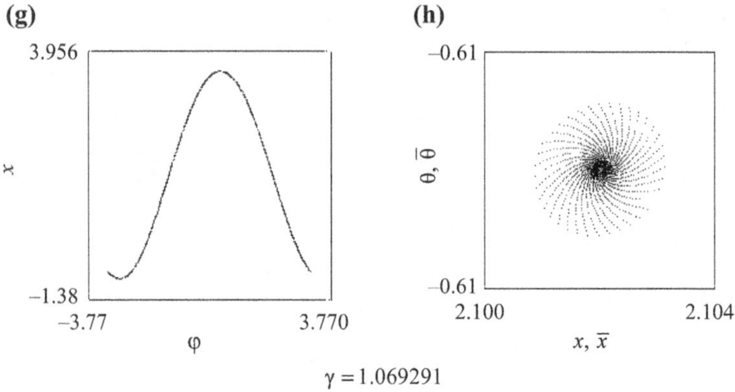

Fig. 4.9 (continued)

left-hand limit cycle stochastically jumps into the vicinity of post-resonance right-hand limit cycle, and further also stochastically goes back to the pre-resonance limit cycle. Further, this scenario is repeated on and on with a random period. During the transition of the system from one limit cycle to another limit cycle, the value of a twist angle of the shaft may reach unacceptably critical values. It is evident that such a dynamical regime affecting both the shaft and the motor is quite unfavorable and may even lead to catastrophic failures.

For the increased values of γ, the strange attractor, first, becomes symmetric (see Fig. 4.9e, f), and then becomes asymmetric. Further, an unstable torus is born from the right-hand limit cycle, while the latter becomes stable, see Fig. 4.9g, h. For a further increase of γ, the invariant curve (in the space of the point mapping) "gets stuck" in the separatrix loop of a saddle, while the fixed point becomes globally stable, i.e. the post-resonance periodic dynamical regime is stable. One can say that for the parameters that correspond to small values of Λ, the Lorenz mechanism of origination and disappearance of the dynamical chaos occurs.

Experiment 2 The following parameter set has been used: $d = -0.27508$, $\mu = 0.01$, $b = -0.01875$, $\omega_0 = 0.5$. Note that the parameters of the averaged system are quite "sensitive" to changes of initial parameters, in particular, parameter b. For these parameters, the value of parameter Λ in system (4.23) is large. In this case, the scenario of origination of chaos differs from the one that has been discussed above. Figure 4.10 shows a gallery of phase portraits and Poincaré mappings for the different values of parameter γ, which is varied such that it approaches the resonance frequency from the right-hand side. For $\gamma < 1$, a globally stable limit cycle occurs. Further, this limit cycle loses stability during the origination of a stable torus (direct Andronov-Hopf bifurcation, see Fig. 4.10a). For increased γ, the torus experiences a cascade of pitchfork bifurcations, see Fig. 4.10b–d. In this case, the torus always remains smooth and ergodic. The series of bifurcations ends with origination of the Feigenbaum attractor, see Fig. 4.10e, f. Further, a crisis occurs that consists of an alternation of the Feigenbaum attractor with "remains" of non-symmetrical Lorenz

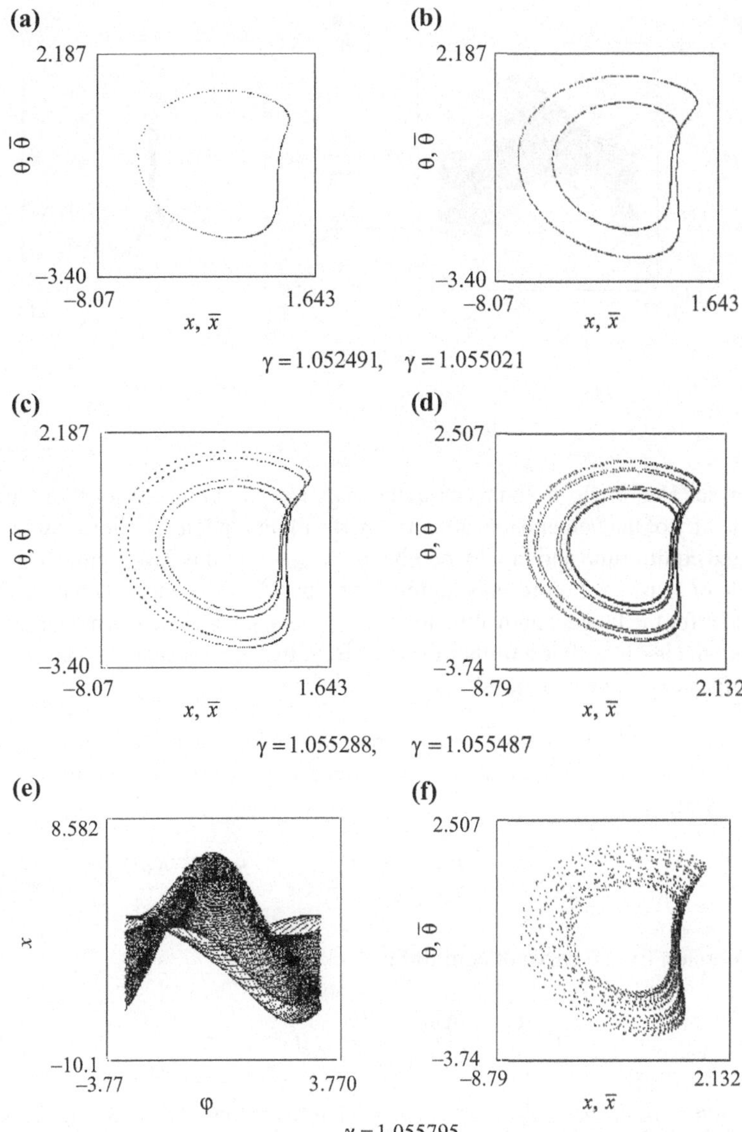

$$\gamma = 1.052491, \quad \gamma = 1.055021$$

$$\gamma = 1.055288, \quad \gamma = 1.055487$$

$$\gamma = 1.055795$$

Fig. 4.10 The gallery of the phase portraits and portraits of the Poincaré mapping for Experiment 2 (continues on the next page)

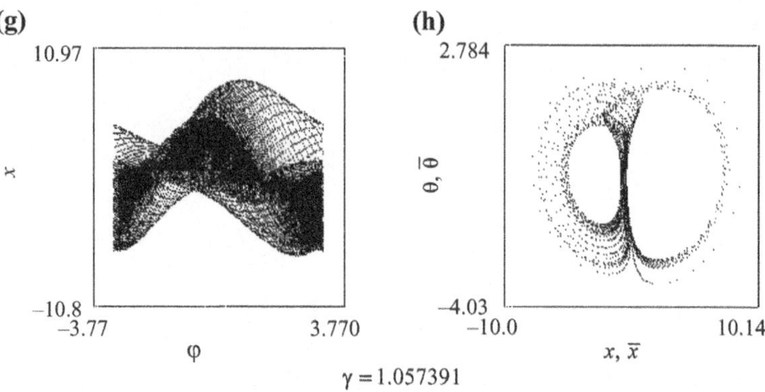

$$\gamma = 1.057391$$

Fig. 4.10 (continued)

attractor, see Fig. 4.10g, h. In this case, the mapping point remains most of the time in the vicinity of the Feigenbaum attractor, performing random ejections towards the right-hand equilibrium and rapidly coming back again. In this case, a time history of the angle of twist $J(t)$ represents quite a long "pack" with chaotic behavior of the amplitude (for the Feigenbaum attractor, the degree of chaotic state is not large) with rare random ejections. For a further increase of γ, the same scenario as described in the previous experiment occurs.

Now let us study qualitative forms of rotation and resonance characteristics for torsional vibrations of the shaft. Let us rewrite expressions for equilibria of the averaged system:

$$x_0 = \omega, \quad y_0 = \frac{R\omega - \Lambda\omega^2}{1 + \omega^2}, \quad z_0 = \frac{R\omega^2 + \Lambda\omega}{1 + \omega^2} \tag{4.24}$$

and expression for the curve of equilibria:

$$f = \omega + \frac{\Lambda\omega^2 - R\omega}{1 + \omega^2}, \quad f = \rho/\sigma. \tag{4.25}$$

Rotation characteristic. Similarly to how it has been done in Sect. 3.2, the function $\rho = \rho(x^*(t, t_0))$, will be considered as a rotation characteristic that has all of the same properties as the original rotation characteristic.

Resonance characteristic (see definition in Sect. 4.1). Let us obtain the expression for the resonance characteristic of torsional vibrations.

From the first equation of system (4.21), we obtain $J - \frac{d}{\omega_0} = \sqrt{\theta^2 + \eta^2}\sin(\varphi + \varphi_0)$, i.e. the expression for the amplitude has the form $A = \sqrt{\theta^2 + \eta^2}$. On the other hand, from formulae (4.22) we obtain $(y + \Lambda)^2 + (z - R)^2 = \frac{b_1^2 b_4^2}{b_3^2 + b_2^2}(\theta^2 + \eta^2)$, i.e. $A = \frac{\sqrt{b_3^2 + b_2^2}}{b_1 b_4} \times \sqrt{(y^* + \lambda)^2 + (z^* - R)^2}$, where

$y^*(t, t_0)$, $z^*(t, t_0)$ is a stationary solution of the averaged system (4.23), which is realized for a given initial condition.

For equilibria of system (4.24) (limit cycles of original system (4.20)), we obtain $(y^* + \Lambda)^2 + (z^* - R)^2 = \frac{R^2 + \lambda^2}{1 + \omega^2}$, i.e. $A = \frac{\sqrt{b_3^2 + b_2^2}\sqrt{R^2 + \lambda^2}}{b_1 b_4} \frac{1}{\sqrt{1 + \omega^2}}$. Since we are interested in qualitative forms of the resonance characteristic, let us consider its normalized form:

$$A^*(\rho) = \frac{1}{\sqrt{1 + (\omega(\rho))^2}}. \tag{4.26}$$

Equations (4.25) and (4.26) represent the rotation characteristic in a parametric form (ω is such a parameter).

Figures 4.11, 4.12 and 4.13 show galleries of rotation and resonance characteristics for different values of parameters Λ and R.

For the curves shown in Fig. 4.11, $\Lambda = 0$. Despite that this case has no physical sense, it can be considered as an approximation for small values of parameter Λ (see Experiment 1). Stable branches are indicated using bold lines, while unstable ones are indicated by dashed lines.

For $R < 1$, the rotation characteristics represent one-one curves and dynamics of the coupled "motor-shaft" system are similar to those for a linear system: the resonance characteristic is a one-one curve, see Fig. 4.11a, b.

For $R > 1$, the rotation characteristic has a hysteretic loop. In this case, the frequency jumps occur from the extrema of the rotation characteristic (to the right during the startup process and to the left during the shutdown), while jumps in amplitudes of torsional vibrations always occur downwards from the points of the rotation characteristic with the vertical tangent $\left(\frac{d\rho}{dA} = 0\right)$.

For the increased values of parameter R, conditions of stability of equilibria for the points of the positively sloped branches of rotation characteristic are not fulfilled anymore (opposite Andronov-Hopf bifurcation), which leads to the decrease of the depth of hysteresis ("cropping" of the hysteretic loop). The same occurs with the resonance characteristic (see Fig. 4.11c, d). Dynamics of the system remain regular for the whole range of values of parameter ρ.

As soon as R reaches the value of R_c (see Sect. 3.2), the depth of hysteresis of the rotation characteristic that corresponds to equilibria tends to zero, while the resonance characteristic becomes non-hysteretic. In this case, despite equilibria of system (4.23) [limit cycles of system (4.20)], there also exists the Lorenz attractor that has already originated. It exists for $\rho = 0$, and, therefore, also for small values of $|\rho|$). Thus, for one set of initial conditions, a resonance characteristic without hysteresis may be realized and for the other set, the motions of the motor and also torsional vibrations of the shaft become chaotic. For $R = R_c + 0$ and $\rho = 0$, the Lorenz attractor becomes a unique attractor of system (4.23). It means that if a startup process is quasi-static, then a jump of the rotor's frequency occurs not to the stable equilibrium as before, but "on a strange attractor". In this case, the instantaneous frequency of rotation varies chaotically in quite a wide range. The same happens to

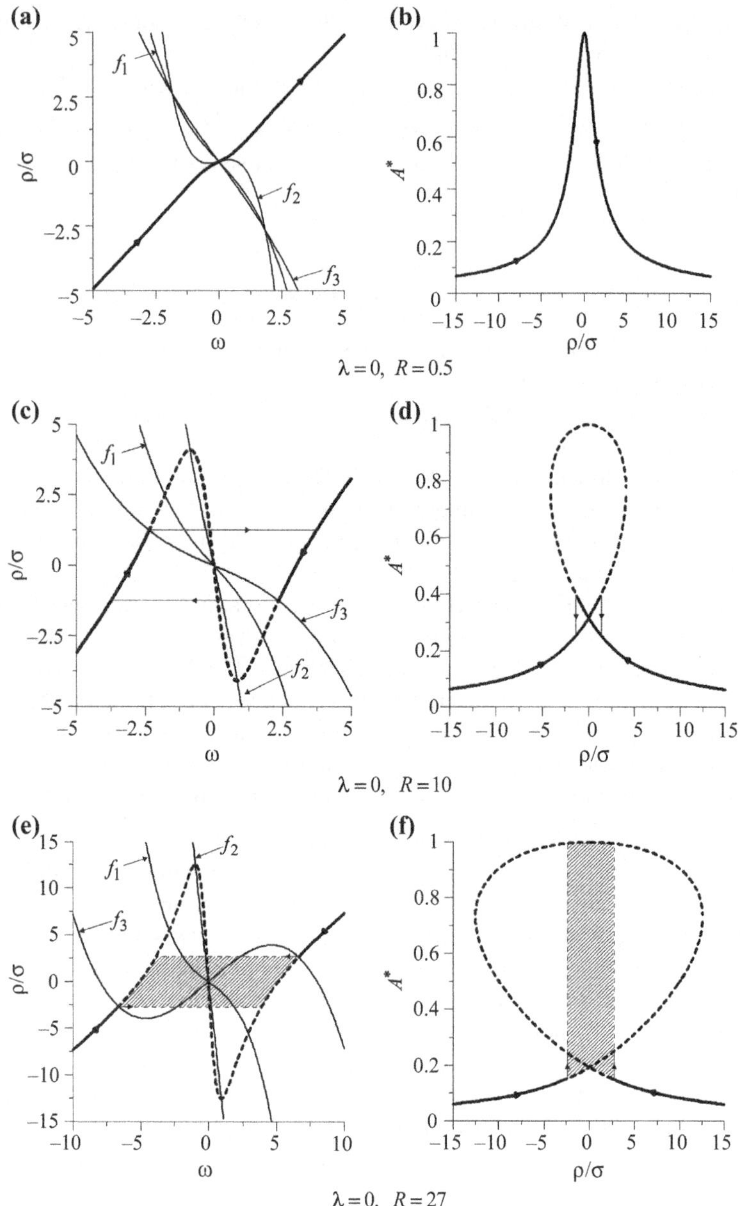

Fig. 4.11 Rotation characteristic and resonance characteristic for $\lambda = 0$ and different values of R (continues on the next page)

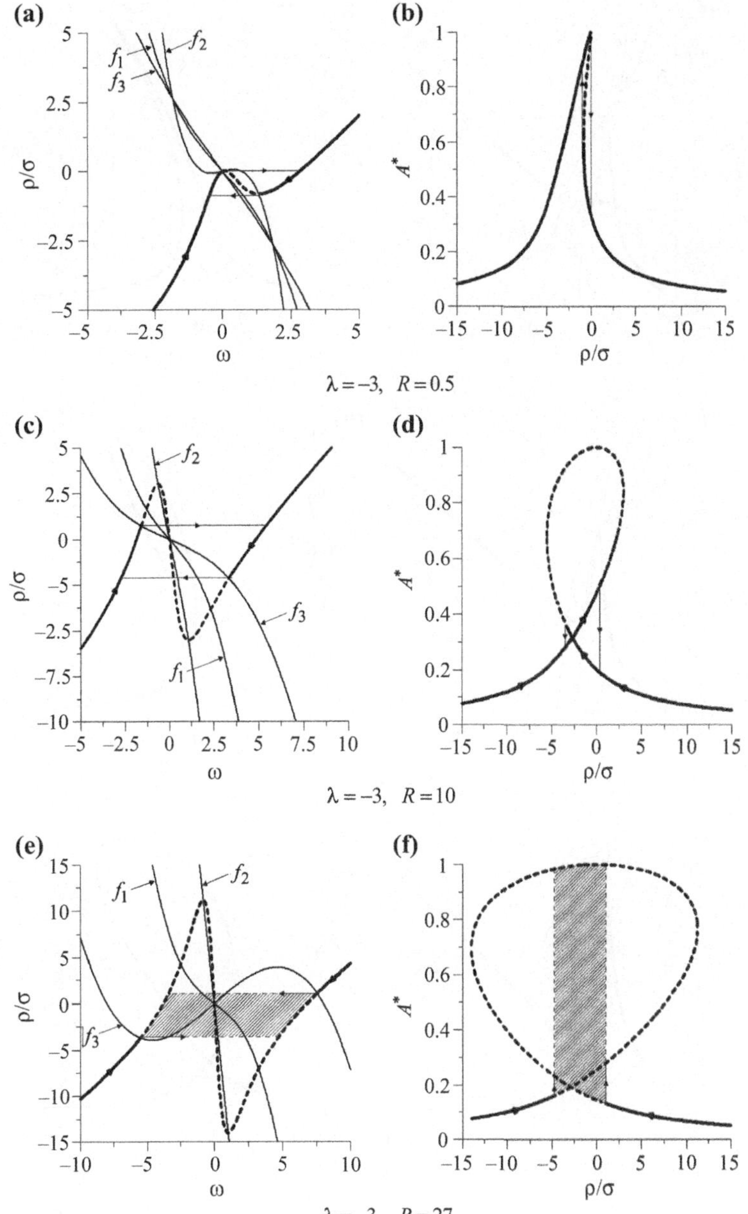

Fig. 4.12 Rotation characteristic and resonance characteristic for $\lambda = -3$ and different values of R (continues on the next page)

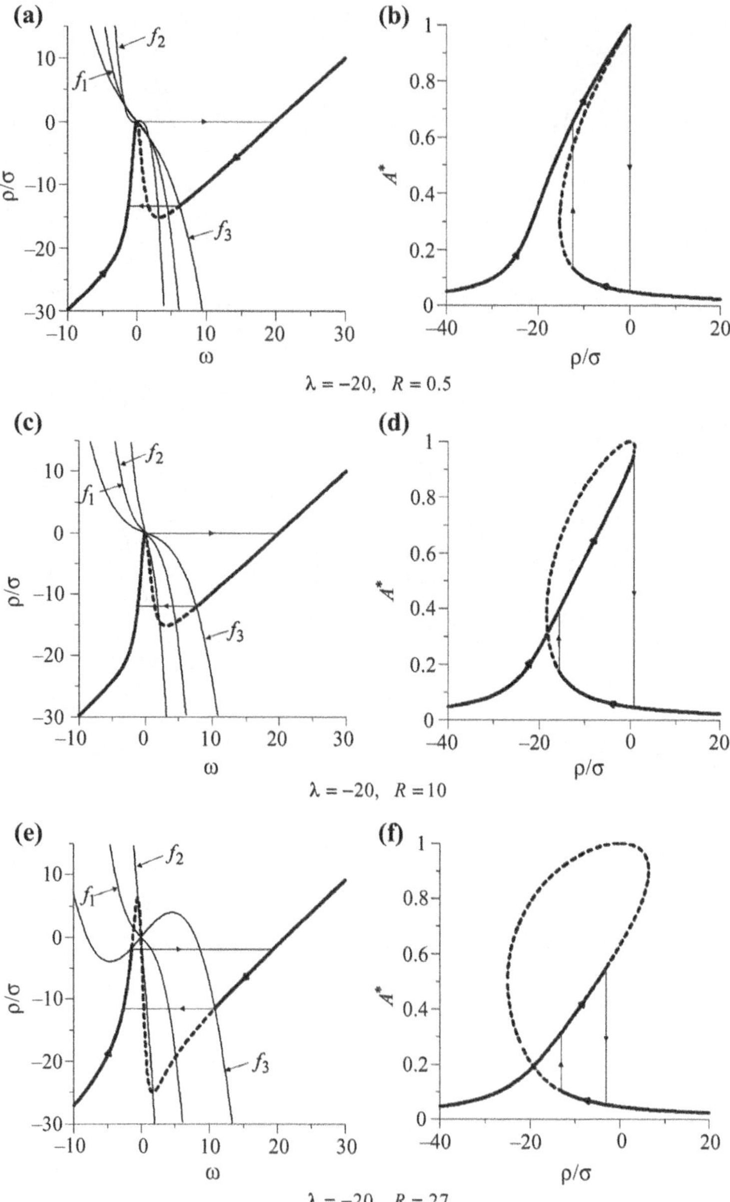

Fig. 4.13 Rotation characteristic and resonance characteristic for $\lambda = -20$ and different values of R (continues on the next page)

the amplitude of torsional vibrations. It seems that such a dynamical regime is rather undesirable for the motor as well as for the shaft. The zone of chaotic vibrations (by parameter ρ), which originates for $R = R_c$, widens as R increases.

Figure 4.11e, f correspond to the case when strange attractors represent unique attractors in a wide range of parameter ρ. During a startup and shutdown process, the system always bypasses a zone of chaotic vibrations. Inside the shaded domains, an effect of the scattering occurs.

The bifurcation scenario of origination and disappearance of the dynamical chaos for parameters of the system that correspond to the small values of Λ is described above in Experiment 1.

Consider the behavior of rotation and resonance characteristics for large values of Λ and various R. Asymmetry of rotation and resonance characteristics, which are not visible for the small values of Λ, become evident for the large values of this parameter. Figure 4.12 shows a gallery of rotation and resonance characteristics for $\Lambda = -3$. In this case, the rotation characteristic is hysteretic including small values of R, while the resonance characteristics bend toward the right-hand side (see Fig. 4.12a, b). As soon as parameter R is increased, the unstable sections of positively sloped branches become larger, as well as parts of resonance curves corresponding to them, see Fig. 4.12c, d. Dynamics of the system are regular.

For the increased values of R, the depth of hysteresis decreases to zero and a zone of scattering of rotation and resonance characteristics that corresponds to the chaotic dynamics of the system occurs, see Fig. 4.12e, f. Inside the shaded domains, a scattering of rotation and resonance characteristics occurs. A bifurcation scenario of origination of the dynamical chaos during the startup process and during the approach to the natural frequency of torsional vibrations has been described above in Experiment 2.

As the experiment shows, for any given $R > R_c$, zones of uncertainty for both characteristics become smaller for increased $|\Lambda|$, and disappear for some value $|\Lambda^*|$. In other words, for all "moderate" values of parameter R and large values of $|\Lambda|$, the dynamics of the system are regular in the whole range of variation of parameter ρ. Figure 4.13 shows a gallery of rotation and resonance characteristics for $\Lambda = -20$ and various R. In all cases, equilibria of system (4.23) corresponding to positively sloped branches of the rotation characteristic lose stability via the opposite Andronov-Hopf bifurcation. At the same time, unstable limit cycles originating from separatrix loops of saddles get stuck in equilibria (this corresponds to the negatively sloped branches of the rotation characteristic). In all cases, the dynamics of the system are regular and a classical Sommerfeld effect occurs.

References

1. Sommerfeld, A.: VDI **18** (1904)
2. Kalischuk, A.K.: Elementary method to study the dynamical properties of systems. J. Theor. Phys. **9**(8), 687–696 (1939). (in Russian)

3. Martyshkin, V.S.: Installation to study the dynamical properties of building materials, in the book "Dynamical properties of building materials". Stroyizdat, Moscow (1940). (in Russian)
4. Blekhman, I.I.: Handbook "Self-synchronization of vibrators of some vibrational machines". Eng. Collect. **16**, 49–72 (1953). (in Russian)
5. Kononenko, V.O.: Vibrating Systems with a Limited Power Supply. Iliffe Books, London (1969)
6. Filippov, A.P.: Vibrations of Deformable Systems. Mashinostroenie, Moscow (1970). (in Russian)
7. Mintz, R.M.: Study of the trajectories of a system of three differential equations at the infinity. Collection of papers in memory of A. A. Andronov, M., izd. AN SSSR (1955) (in Russian)
8. Verichev, N.N., Verichev, S.N.: Damping of ESP lateral vibrations using modulation of motor speed. Patent WO2009096806 (2008)
9. Verichev, N.N., Verichev, S.N., Erofeev, V.I.: Damping of bending vibrations of rotating shaft. Bull. Mach. Build. **8**, 26–30 (2012). (in Russian)
10. Verichev, N.N.: Chaotic torsional vibration of imbalanced shaft driven by a limited power supply. J. Sound Vib. **331**(2), 384–393 (2012)
11. Likharev, K.K.: Dynamics of Josephson Junctions and Circuits. Gordon and Breach, New York (1986)
12. Verichev, N.N.: Study of systems with Josephson junctions by the method of a spinning phase. Radiotechnika i Elektronika **31**(11), 2267–2274 (1986). (in Russian)

Chapter 5
Synchronization in Homogeneous Lattices

The interest in lattices of dynamical systems (oscillators, rotators) has two aspects. On one hand, lattices represent adequate discrete models of active continuous media [1, 2], for instance, to study the mechanisms of turbulence development [3–5]. On the other hand, they represent various physical and biological systems that have an array or network structure exemplified, for instance, by systems of phase synchronization [6], neural ensembles [7, 8], neuron-like information processing networks [9–11], etc.

This chapter studies stability of synchronization in homogeneous lattices of dynamical systems, which defines their spatially-homogeneous state. We consider diffusive-coupled lattices of different geometrical dimensions [12].

5.1 Synchronization in Lattices of Dynamical Systems. General Information

Consider a homogeneous lattice of diffusive-coupled dynamical systems of the form

$$\dot{\mathbf{X}}_i = \mathbf{F}(\mathbf{X}_i, t) + \varepsilon \mathbf{C}(\mathbf{X}_{i-1} - 2\mathbf{X}_i + \mathbf{X}_{i+1}),$$
$$i = \overline{1, N}. \tag{5.1}$$

An individual system in (5.1) is governed by the equation

$$\dot{\mathbf{X}} = \mathbf{F}(\mathbf{X}, t),$$
$$\mathbf{X} = (x_1, x_2, \dots, x_m)^T, \ x \in R^1, \ \mathbf{F}(\mathbf{X}) : R^m \to R^m. \tag{5.2}$$

In principle, any model of physical systems discussed in the previous sections can act here as an individual dynamical system.

© Springer Nature Switzerland AG 2020

109

N. Verichev et al., *Chaos, Synchronization and Structures in Dynamics of Systems with Cylindrical Phase Space*, Understanding Complex Systems, https://doi.org/10.1007/978-3-030-36103-7_5

We will assume that the dynamical properties of system (5.2) are known. In particular, we assume that in the extended phase space $G(\mathbf{X}, t)$ there exists attractor $A(1)$ with maximum Lyapunov exponent $\lambda(1)$. The latter represents the maximum exponent of the equation in variations

$$\dot{\mathbf{U}} = \mathbf{F}'(\boldsymbol{\xi}(t))\mathbf{U},$$

where $\xi(t) \in A(1)$, $F'(\mathbf{X}) = \frac{\partial \mathbf{F}(\mathbf{X})}{\partial \mathbf{X}}$ is the Jacobi matrix.

Note that for assumptions made the following is valid: $\|\mathbf{U}\| = D\exp\lambda(1)t$, $D = \text{const}$.

Matrix C in Eq. (5.1) defines a structure of couplings of partial systems, while ε represents a scalar parameter.

In this case, if one of the scalar variables in (5.2) is cyclic $x = \varphi(\mathrm{mod}2\pi)$, then the system could be considered as a generalized rotator. This name is justified by the fact that all aforementioned models can be reduced to the form (5.2) as long as two-dimensional rotators are considered along with external forces and loads. We assume that function $\mathbf{F}(\mathbf{X}, t)$ is cyclic in time.

For the boundary conditions of the form $\mathbf{X}_0 \equiv \mathbf{X}_1$, $\mathbf{X}_N \equiv \mathbf{X}_{N+1}$, system (5.1) represents a chain, while for $\mathbf{X}_0 \equiv \mathbf{X}_N$, $\mathbf{X}_1 \equiv \mathbf{X}_{N+1}$, it represents a ring.

It will be seen from further discussion that the synchronization, which represents our main interest, is closely related to the existence of so-called integral manifolds representing singular surfaces in the phase space of coupled systems filled with trajectories. This mathematical object is such that if the initial conditions of the system are specified on the integral surface, then the affix that moves along the corresponding phase trajectory never leaves this surface.

It follows directly from system (5.1) that hyperplane $M_0 = \{\mathbf{X}_1 = \mathbf{X}_2 = \ldots = \mathbf{X}_{N-1} = \mathbf{X}_N\}$. represents its integral manifold. We can draw this conclusion by solving equation $\mathbf{X}_{i-1} - 2\mathbf{X}_i + \mathbf{X}_{i+1} = 0$. The traces of this hyperplane on coordinate planes of the corresponding variables of different partial systems are the bisectors of the first and third coordinate angles. We also note that the geometric dimension of the hyperplane coincides with the dimension of the phase space of the individual system.

Our special interest in this manifold will become clear from the following considerations. Let us imagine that the initial conditions for system (5.1) are given at M_0. In this case, as it directly follows from (5.1), the coupled system falls into N separate systems of the form (5.2). Apart from that, the motions of these independent systems will be synchronized, since, in principle, these are the motions of identical systems with identical initial conditions. In other words, the synchronization of interest corresponds to the motion of a phase point along trajectories located on the integral manifold of system (5.1). The question that remains is: what is the nature of these trajectories? The answer is quite simple: since projections of these trajectories onto the phase spaces of individual systems (partial phase portraits) are the trajectories of partial systems, then trajectories on the hyperplane M_0 are induced by the same system of the form (5.2).

Of course, setting up exact initial conditions (on an integral manifold) lays behind the scope of a real experiment. To solve the problem of synchronization, the problem of stability of an integral manifold must be solved.

Assume that the integral manifold is stable. In this case, the process of synchronization can be described as follows. For any initial condition from the neighborhood of the manifold (at least from a small neighborhood of it), the affix runs to the surface of M_0 and, after some time, starts moving almost along the trajectory of attractor $A(1)$ of system (5.2) defined on this manifold. In this case, all of the partial phase points in the spaces of individual systems [projections of the phase point of system (5.1)] move in mutual synchronism along the corresponding trajectories of their partial phase portraits, i.e. along identical trajectories of identical attractors $A(1)$. In other words, there is a stable regime of mutual synchronization of interacting subjects in the lattice (5.1).

Synchronization in a chain. Consider the local stability of synchronization as a local stability of a "part" of the manifold that contains attractor $A(1)$. It can be established that, with respect to transversals $\mathbf{U} = \left(\mathbf{U}_1, \ \mathbf{U}_2, \ .., \ \mathbf{U}_p\right)$ of manifold M_0, where $\mathbf{U}_k = \mathbf{X}_k - \mathbf{X}_{k+1}, k = \overline{1, p}, p = N - 1$, linearized system (5.1) can be reduced to one equation of the form

$$\dot{\mathbf{U}} = \left(\mathbf{I}_p \otimes \mathbf{J}_m(t) - \varepsilon \mathbf{D}_p \otimes \mathbf{C}\right)\mathbf{U}. \tag{5.3}$$

Here $J_m(\xi(t))$ is the Jacobi matrix of an elementary oscillator, $\xi(t) \in A(1)$, $\otimes$ is the symbol of the direct (Kronecker) product of matrices, $\mathbf{I}_p$ is the unit matrix of the corresponding order, $\mathbf{D}_p$ is the symmetric matrix of the form

$$\mathbf{D}_p = \begin{pmatrix} 2 & -1 & 0 & 0 & & & & & \\ -1 & 2 & -1 & 0 & & & \mathbf{0} & & \\ 0 & -1 & 2 & -1 & & & & & \\ 0 & 0 & -1 & 2 & & & & & \\ & & & & \ddots & & & & \\ & & & & & 2 & -1 & 0 & 0 \\ & \mathbf{0} & & & & -1 & 2 & -1 & 0 \\ & & & & & 0 & -1 & 2 & -1 \\ & & & & & 0 & 0 & -1 & 2 \end{pmatrix}.$$

Stable isochronous synchronization corresponds to the stable solution $\mathbf{U} = 0$ of Eq. (5.3). This equation has one property that radically simplifies solution of the problem that we formulate in the form of a lemma (see also [13]).

Lemma 5.1 *Suppose $\lambda_{\min} = \lambda_1 < \lambda_2 < \lambda_3 < \ldots < \lambda_{p-1} < \lambda_p = \lambda_{\max}$ is the spectrum of eigenvalues of constant matrix $\mathbf{D}_p$. In this case, Eq. (5.3), as a result of a nondegenerate transformation $\mathbf{U} = \mathbf{SV}$, is reduced to the form*

$$\dot{\mathbf{V}}_j = \left(\mathbf{J}_m(t) - \varepsilon\lambda_j\mathbf{C}\right)\mathbf{V}_j,$$

$$j = \overline{1, p}. \tag{5.4}$$

Proof Suppose $\mathbf{S}_0$ is the transform matrix of matrix $\mathbf{D}_p$, while $\mathbf{S}$ is the transform matrix of matrix $\mathbf{D}_p \otimes \mathbf{C}$, i.e. $\mathbf{S}_0^{-1}\mathbf{D}_p\mathbf{S}_0 = \mathbf{D}_0 = \mathrm{diag}(\lambda_1, \lambda_2, \ldots, \lambda_p)$. It is known that the roots of matrix $\mathbf{D}_p \otimes \mathbf{C}$ are all possible products of the roots of matrix-factors: $\lambda_1 c_1, \lambda_1 c_2, \ldots, \lambda_1 c_m; \lambda_2 c_1, \lambda_2 c_2, \ldots, \lambda_2 c_m; \ldots; \lambda_p c_1, \lambda_p c_2, \ldots, \lambda_p c_m$, i.e. $\mathbf{S}^{-1}\left(\mathbf{D}_p \otimes \mathbf{C}\right)\mathbf{S} = \mathbf{D}_0 \otimes \mathbf{C}$. Let us show that $\mathbf{S} = \mathbf{S}_0 \otimes \mathbf{I}_m$. Applying the properties of direct product of matrices [14], we obtain

$$\mathbf{S}^{-1}\left(\mathbf{D}_p \otimes \mathbf{C}\right)\mathbf{S} = \mathbf{S}^{-1}\left(\mathbf{S}_0\mathbf{D}_0\mathbf{S}_0^{-1} \otimes \mathbf{I}_m\mathbf{C}\mathbf{I}_m^{-1}\right)\mathbf{S}$$
$$= \mathbf{S}^{-1}(\mathbf{S}_0 \otimes \mathbf{I}_m)(\mathbf{D}_0 \otimes \mathbf{C})\left(\mathbf{S}_0^{-1} \otimes \mathbf{I}_m\right)\mathbf{S} = \mathbf{D}_0 \otimes \mathbf{C}.$$

From here it follows that $\mathbf{S}^{-1}(\mathbf{S}_0 \otimes \mathbf{I}_m) = \mathbf{I}_{mp}$ and $\mathbf{S} = \mathbf{S}_0 \otimes \mathbf{I}_m$.

Let us perform the transformation $\mathbf{U} = \mathbf{SV}$. in Eq. (5.3). As a result, we obtain an equation of the form

$$\dot{\mathbf{V}} = \mathbf{S}^{-1}\left(\mathbf{I}_p \otimes \mathbf{J}_m\right)\mathbf{SV} - \varepsilon\mathbf{S}^{-1}\left(\mathbf{D}_p \otimes \mathbf{C}\right)\mathbf{SV}.$$

Let us show that matrix $\mathbf{S}^{-1}\left(\mathbf{I}_p \otimes \mathbf{J}_m\right)\mathbf{S}$ is cell-diagonal. Using the properties of the direct product of matrices, we obtain:

$$\mathbf{S}^{-1}\left(\mathbf{I}_p \otimes \mathbf{J}_m\right)\mathbf{S} = \left(\mathbf{S}_0^{-1} \otimes \mathbf{I}_m\right)\left(\mathbf{I}_p \otimes \mathbf{J}_m\right)(\mathbf{S}_0 \otimes \mathbf{I}_m)$$
$$= \left(\mathbf{S}_0^{-1}\mathbf{I}_p\mathbf{S}_0\right) \otimes (\mathbf{I}_m\mathbf{J}_m\mathbf{I}_m) = \mathbf{I}_p \otimes \mathbf{J}_m = diag(\mathbf{J}_m, \mathbf{J}_m, \ldots, \mathbf{J}_m).$$

Taking into account all transformations with respect to the coordinates of vector $\mathbf{V}$, we obtain system of Eq. (5.4), equivalent to Eq. (5.3).

We note that the transformation of Eq. (5.3) to Eq. (5.4) is analogous by sense to the reduction of a system of linear and linearly coupled oscillators to normal vibrations (normal form).

On the basis of this lemma, we formulate the following theorem on the local stability of synchronization.

Theorem 5.1 *Suppose $\lambda(1)$ is the maximal exponent of attractor $A(1)$ of a partial oscillator, while $\lambda_{\min}$, $\lambda_{\max}$ are, respectively, the minimum and maximum eigenvalues of matrix $\mathbf{D}_p$. If matrix $\mathbf{C} = \mathbf{I}$, then synchronization in a coupled system of oscillators is stable for $\varepsilon > \lambda(1)/\lambda_{\min}$ and unstable for $\varepsilon > \lambda(1)/\lambda_{\min}$ and unstable for the opposite inequality. For $\varepsilon < \lambda(1)/\lambda_{\min}$, it is unstable in all directions transverse to manifold M_0. For intermediate values of the parameter of coupling, the synchronization is multistable. For the other matrix, matrix $\mathbf{C}$, the same inequalities are valid, but for a change $\lambda(1) \rightarrow \varepsilon^*(\lambda(1))$, where the value of ε^* depends on this matrix.*

Proof Suppose that $\mathbf{C} = \mathbf{I}$. Let us perform a change of the variables in Eq. (5.4): $\mathbf{V}_j = e^{-\varepsilon\lambda_j t}\mathbf{W}_j$, which results in a system of identical equations of the form $\dot{\mathbf{W}}_j = \mathbf{J}_m(t)\mathbf{W}_j$. By their form, each of these equations is an equation in variation with respect to the solution corresponding to a trajectory of the attractor of a single oscillator, i.e. $\|\mathbf{W}_j\| = Ce^{\lambda(1)t}$, where C is a certain constant. As a result of the change of the variables, we obtain $\|\mathbf{V}_j\| = Ce^{(\lambda(1)-\varepsilon\lambda_j)t}$. From here, it follows that if $\varepsilon > \lambda(1)/\lambda_j$, then $\|\mathbf{V}_j\| \to 0$. for $t \to \infty$, as well as $\|\mathbf{V}_j\| \to \infty$ for $t \to \infty$, if $\varepsilon < \lambda(1)/\lambda_j$. It means that if $\varepsilon > \lambda(1)/\lambda_{\min}$, then all Eq. (5.4) are stable, and if $\varepsilon < \lambda(1)/\lambda_{\min}$, then all equations of (5.4) are unstable. According to the lemma, the conditions of stability and instability are also transferred to Eq. (5.3). For intermediate values of parameter ε, a number of equations is stable and the rest are unstable; solution $\mathbf{V} = 0$, and, therefore, also $\mathbf{U} = 0$ have a saddle type, which corresponds to multistable, or, in other words, to partial synchronization.

Function $\varepsilon^*(\lambda(1))$ equals zero for $\lambda(1) \leq 0$ and grows for $\lambda(1) > 0$. Apart from that, it depends on the structure of coupling of the oscillators representing elements c_i of matrix $\mathbf{C}$. In particular, if $\mathbf{C} = \mathbf{I}$, then $\varepsilon^*(\lambda(1)) = \lambda(1)$.

Let us analyze conditions of stability of synchronization.

Determination of the eigenvalues of matrix $\mathbf{D}_p$ [15] for the minimum root results in: $\lambda_{\min} = 4\sin^2\frac{\pi}{2N}$. After this, the condition of stability takes the form

$$\varepsilon > \varepsilon^*(\lambda(1))/\left(4\sin^2\frac{\pi}{2N}\right) \tag{5.5}$$

Let us assume that the attractor of an individual system is regular (such as with a stable limit cycle or torus), i.e. $\lambda(1) < 0$, $\varepsilon^*(\lambda(1)) = 0$. As one can see from (5.5), in this case, synchronization in the chain is stable for any value of the parameter of coupling, including arbitrary, small values. And this is the case for chains with any number of elements.

A completely different situation arises in the case when the attractor of an individual system is chaotic. In this case, $\lambda(1) > 0$, $\varepsilon^*(\lambda(1)) > 0$, and, therefore, the right-hand side of (5.5) establishes a threshold value of the parameter of coupling that depends on the degree of randomness of the attractor, the measure of which is $\lambda(1)$, as well as on the number of dynamical systems N in the chain. For a fixed value of ε, the number of systems in the chain for which a chaotic synchronization can take place is defined by an inequality of the form

$$N \leq N_0 = \left[\frac{\pi}{\arccos\nu}\right], \quad \nu = 1 - \frac{\varepsilon^*(\lambda(1))}{2\varepsilon}.$$

In particular, if $N \gg 1$, then the following asymptotic formulae are valid: $\varepsilon > \pi^{-2}N^2\varepsilon^*(\lambda(1))$, $N \leq N_0 = \left[\pi\sqrt{\varepsilon/\varepsilon^*}\right]$.

Synchronization in the ring. Consider the stability of the synchronization in the ring. Taking into account boundary conditions $\mathbf{X}_0 \equiv \mathbf{X}_N$, $\mathbf{X}_1 \equiv \mathbf{X}_{N+1}$ with respect to variables $\mathbf{U}_k = \mathbf{X}_k - \mathbf{X}_{k+1}$, $k = \overline{1, N-1}$ representing transversals to the "part" of the manifold that contains the chaotic attractor, we obtain an equation of the same form as (5.3), with matrix $\mathbf{D}_p$, $p = N - 1$ of the form

$$
\mathbf{D}_p =
\begin{pmatrix}
3 & 0 & 1 & 1 & \ldots & 1 & 1 & 1 & 1 \\
-1 & 2 & -1 & 0 & \ldots & 0 & 0 & 0 & 0 \\
0 & -1 & 2 & -1 & \ldots & 0 & 0 & 0 & 0 \\
0 & 0 & -1 & 2 & \ldots & 0 & 0 & 0 & 0 \\
\vdots & \vdots & \vdots & \vdots & \ddots & \vdots & \vdots & \vdots & \vdots \\
0 & 0 & 0 & 0 & \ldots & 2 & -1 & 0 & 0 \\
0 & 0 & 0 & 0 & \ldots & -1 & 2 & -1 & 0 \\
0 & 0 & 0 & 0 & \ldots & 0 & -1 & 2 & -1 \\
1 & 1 & 1 & 1 & \ldots & 1 & 1 & 0 & 3
\end{pmatrix}.
$$

Eigenvalues of matrix $\mathbf{D}_p$ are expressed by formula $\lambda_j = 4\sin^2 \pi j/N$, $j = \overline{1, N-1}$, from which it follows that $\lambda_{\min} = 4\sin^2 \frac{\pi}{N}$, and $\lambda_{\max} = 4\sin^2 \frac{\pi j^*}{N}$, where $j^* = \frac{N}{2}$, if N is an even number, and $j^* = \frac{N+1}{2}$, if N is an odd one.

Thus, isochronous synchronization in a ring of dynamical systems is stable if $\varepsilon > \varepsilon^*/4\sin^2 \frac{\pi}{N}$, while for $\varepsilon < \varepsilon^*/4\sin^2 \frac{\pi j^*}{N}$, it is unstable.

Synchronization in a two-dimensional lattice. Let us consider the stability of the synchronization of dynamical systems in a two-dimensional rectangular lattice modeled by differential equations of the form

$$
\begin{aligned}
\dot{\mathbf{X}}_{ij} &= \mathbf{F}(\mathbf{X}_{ij}) - \varepsilon\mathbf{C}(4\mathbf{X}_{ij} - \mathbf{X}_{i-1j} - \mathbf{X}_{i+1j} - \mathbf{X}_{ij-1} - \mathbf{X}_{ij+1}), \\
&\quad i = 1, 2, \ldots N_1, \ \ j = 1, 2, \ldots N_2
\end{aligned}
\tag{5.6}
$$

with boundary conditions: $\mathbf{X}_{0j} = \mathbf{X}_{1j}$, $\mathbf{X}_{N_1+1j} = \mathbf{X}_{N_1j}$, $\mathbf{X}_{i0} = \mathbf{X}_{i1}$, $\mathbf{X}_{iN_2+1} = \mathbf{X}_{iN_2}$.

Using the chain as an example, we have seen that the study of the stability of synchronization is ultimately reduced to the study of the eigenvalues of matrix $\mathbf{D}_p$, which is responsible for couplings of individual systems. Obviously, the same should take place in the case of Eq. (5.6). The problem of finding the corresponding matrix can be solved by a fusion of a two-dimensional lattice from the corresponding chains, using them as "elements" of the lattice.

Consider a rectangular lattice as a "parallel" coupling of chains. To do so, we equip the vectors of the elementary oscillator in (5.1) with a second subscript: we introduce vectors $\mathbf{Y}_j = \mathrm{col}(\mathbf{X}_{1j}, \mathbf{X}_{2j}, \ldots, \mathbf{X}_{N_1j})$, $\Phi(Y_j) = \mathrm{col}(\mathbf{F}(\mathbf{X}_{1j}), \mathbf{F}(\mathbf{X}_{2j}), \ldots, \mathbf{F}(\mathbf{X}_{N_1j}))$ and rewrite an arbitrary chain [system (5.1)] with one equation:

$$
\dot{\mathbf{Y}}_j = \Phi(\mathbf{Y}_j) - \varepsilon(\mathbf{B}_{N_1} \otimes \mathbf{C})\mathbf{Y}_j,
\tag{5.7}
$$

Where $\mathbf{B}_{N_1}$ is a diffusion matrix of the form

$$\mathbf{B}_{N_1} = \begin{Vmatrix} 1 & -1 & 0 & \cdots & \cdots & 0 & 0 \\ -1 & 2 & -1 & \ddots & \cdots & 0 & 0 \\ 0 & -1 & 2 & \ddots & \ddots & 0 & 0 \\ \vdots & \ddots & \ddots & \ddots & \ddots & \ddots & \vdots \\ \vdots & \vdots & \ddots & \ddots & 2 & -1 & 0 \\ 0 & 0 & 0 & \ddots & -1 & 2 & -1 \\ 0 & 0 & 0 & \cdots & 0 & -1 & 1 \end{Vmatrix}.$$

Since, for a parallel coupling of the chains, all elements of vector $\mathbf{Y}_j$ are coupled to the corresponding elements of the neighboring vectors $\mathbf{Y}_{j-1}$ and $\mathbf{Y}_{j+1}$, the matrix of the fusion of the lattice from the elements of (5.7) represents a unit matrix $\mathbf{I}_{N_1}$. Taking the aforesaid parts, as well as Eq. (5.7), the equations of the lattice take the form

$$\dot{\mathbf{Y}}_j = \boldsymbol{\Phi}(\mathbf{Y}_j) - \varepsilon(\mathbf{B}_{N_1} \otimes \mathbf{C})\mathbf{Y}_j - \varepsilon(\mathbf{I}_{N_1} \otimes \mathbf{C})(-\mathbf{Y}_{j-1} + 2\mathbf{Y}_j - \mathbf{Y}_{j+1}),$$
$$j = 1, 2, \ldots N_2. \tag{5.8}$$

The boundary conditions in system (5.8) have the form $\mathbf{Y}_0 = \mathbf{Y}_1$, $\mathbf{Y}_{N_2} = \mathbf{Y}_{N_2+1}$.

Performing matrix operations and writing this system with respect to elementary oscillators, one can verify the equivalence of systems (5.8) and (5.6). Let us pay attention to the similarity of the forms of Eqs. (5.8) and (5.1).

In contrast to the previous section, where the linearization was carried out for variables $\mathbf{U}_i = \mathbf{X}_i - \mathbf{X}_{i+1}$ (transversals of the manifold), here we will linearize Eq. (5.8) in a standard way, for variables $\mathbf{U}_{ij} = \mathbf{X}_{ij} - \boldsymbol{\xi}(t)$ in the vicinity of solution $\mathbf{X}_{ij} = \boldsymbol{\xi}(t) \in A(1)$.

After convolving the linearized system into one equation, we obtain

$$\dot{\mathbf{U}} = \mathbf{I}_{N_1 N_2} \otimes \mathbf{J}_m(\xi)\mathbf{U} - \varepsilon(\mathbf{I}_{N_2} \otimes \mathbf{B}_{N_1} + \mathbf{B}_{N_2} \otimes \mathbf{I}_{N_1}) \otimes \mathbf{CU}. \tag{5.9}$$

We note that the method of linearization does not fundamentally change the essence of the matter. The formal difference is that, in this case, the dimension of matrix $\mathbf{D}_p = \mathbf{I}_{N_2} \otimes \mathbf{B}_{N_1} + \mathbf{B}_{N_2} \otimes \mathbf{I}_{N_1}$ is one unit higher than the one that would be obtained from the first case, and in its spectrum there is one zero root. All other eigenvalues are positive and coincide with the eigenvalues of the matrix obtained in the first case (the matrices are related by equivalent transformations). Without going into much detail, in what follows, we will not take the zero root into account.

As one can see, Eq. (5.9) falls under the above formulated lemma. Consequently, the stability of the synchronization is determined by the roots of matrix $\mathbf{D}_p = \mathbf{I}_{N_2} \otimes \mathbf{B}_{N_1} + \mathbf{B}_{N_2} \otimes \mathbf{I}_{N_1}$.

A study of this matrix shows that its minimal eigenvalue is expressed by the formula, which is already familiar to us from the previous section: $\lambda_{\min} =$

$4\sin^2\frac{\pi}{2N}$, where $N = \max\{N_1, N_2\}$. We obtain an interesting and unexpected result: the condition of the stability of synchronization is defined by inequality $\varepsilon > \varepsilon^*(\lambda(1))/(4\sin^2\frac{\pi}{2N})$, coinciding with the condition of stability of synchronization in a chain whose "length" is equal to the greatest "side" of a rectangular lattice of oscillators.

Synchronization in a three-dimensional lattice. We consider the stability of a spatially homogeneous structure in a three-dimensional lattice of dynamical systems governed by equations of the form

$$\dot{\mathbf{X}}_{ijk} = \mathbf{F}(\mathbf{X}_{ijk})$$
$$- \varepsilon\mathbf{C}\big(6\mathbf{X}_{ijk} - \mathbf{X}_{i-1jk} - \mathbf{X}_{i+1jk} - \mathbf{X}_{ij-1k} - \mathbf{X}_{ij+1k} - \mathbf{X}_{ijk-1} - \mathbf{X}_{ijk+1}\big),$$
$$i = 1, 2, \ldots N_1, \quad j = 1, 2, \ldots N_2, \quad k = 1, 2, \ldots N_3. \tag{5.10}$$

To find the desired matrix, we proceed analogously to the two-dimensional case, by considering a three-dimensional lattice in the form of a "parallel" coupling of layers— two-dimensional lattices of the form (5.6). We equip each elementary oscillator in (5.8) by the third index and write the equation of an arbitrary layer [system (5.8)] in the form of a single equation:

$$\dot{\mathbf{Z}}_k = \mathrm{H}(\mathbf{Z}_k) - \varepsilon\big(\mathbf{I}_{N_2} \otimes \mathbf{B}_{N_1} + \mathbf{B}_{N_2} \otimes \mathbf{I}_{N_1}\big) \otimes \mathbf{C}\mathbf{Z}_k. \tag{5.11}$$

Now, by considering the three-dimensional lattice as a "parallel coupling of layers", we rewrite it in the following form:

$$\dot{\mathbf{Z}}_k = \mathrm{H}(\mathbf{Z}_k) - \varepsilon\big(\mathbf{I}_{N_2} \otimes \mathbf{B}_{N_1} + \mathbf{B}_{N_2} \otimes \mathbf{I}_{N_1}\big) \otimes \mathbf{C}\mathbf{Z}_k$$
$$- \varepsilon\big(\mathbf{I}_{N_1 N_2} \otimes \mathbf{C}\big)(-\mathbf{Z}_{k-1} + 2\mathbf{Z}_k - \mathbf{Z}_{k+1}),$$
$$k = 1, 2, \ldots N_3.$$

The boundary conditions have the form: $\mathbf{Z}_0 = \mathbf{Z}_1$, $\mathbf{Z}_{N_3} = \mathbf{Z}_{N_3+1}$.

Linearization of the obtained system in the vicinity of solution $\mathbf{X}_{ijk} = \boldsymbol{\xi}(t) \in A(1)$ and its subsequent reduction to one equation gives the following result:

$$\dot{\mathbf{U}} = \mathbf{I}_{N_1 N_2 N_3} \otimes \mathbf{J}_m(\xi)\mathbf{U}$$
$$- \varepsilon\big(\mathbf{I}_{N_3} \otimes \mathbf{I}_{N_2} \otimes \mathbf{B}_{N_1} + \mathbf{I}_{N_3} \otimes \mathbf{B}_{N_2} \otimes \mathbf{I}_{N_1} + \mathbf{B}_{N_3} \otimes \mathbf{I}_{N_1 N_2}\big) \otimes \mathbf{C}\mathbf{U}.$$

As one can see, this equation has the required form of the equation of the lemma. Consequently, the problem of stability of the synchronization is reduced to determining the eigenvalues of matrix $\mathbf{D}_p = \mathbf{I}_{N_3} \otimes \mathbf{I}_{N_2} \otimes \mathbf{B}_{N_1} + \mathbf{I}_{N_3} \otimes \mathbf{B}_{N_2} \otimes \mathbf{I}_{N_1} + \mathbf{B}_{N_3} \otimes \mathbf{I}_{N_1 N_2}$.

During the study of this matrix, we establish the aforementioned fact: the minimum root of matrix $\mathbf{D}_p$ equals $4\sin^2\frac{\pi}{2N}$, where $N = \max\{N_1, N_2, N_3\}$. Hence, we obtain the synchronization stability condition, which is already known to us: $\varepsilon > \varepsilon^*(\lambda(1))/\big(4\sin^2\frac{\pi}{2N}\big)$, where $N = \max\{N_1, N_2, N_3\}$. The conclusions regarding the influence of parameters of the attractors of an individual dynamical system and the parameters of the lattice on the stability of the synchronization will be the same as in the case of a chain.

5.2 Synchronization of Rotations in a Chain and in a Ring of Diffusive-Coupled Autonomous and Non-autonomous Rotators

Let us study the processes of synchronization in a system of coupled rotators of the form

$$I\ddot{\varphi}_i + \dot{\varphi}_i + \sin \varphi_i = \gamma + \varepsilon(\dot{\varphi}_{i-1} - 2\dot{\varphi}_i + \dot{\varphi}_{i+1}) + A \sin \psi,$$

$$\dot{\psi} = \omega_0,$$

$$i = \overline{1, N} \tag{5.12}$$

With boundary conditions of two types: a) $\varphi_0 \equiv \varphi_1$, $\varphi_N \equiv \varphi_{N+1}$, corresponding to a chain, and b) $\varphi_0 \equiv \varphi_N$, $\varphi_{N+1} \equiv \varphi_1$, corresponding to a ring.

Variables and parameters in system (5.12) are dimensionless, and this system is defined in the toroidal phase space $G(\varphi_1, \ldots, \varphi_N, \psi, \dot{\varphi}_1, \ldots, \dot{\varphi}_N) = T^{N+1} \times R^N$. Parameters $I > 0$, $\gamma > 0$ have a sense of the moment of inertia and constant torque of rotator, respectively; A, ω_0 are the amplitude and frequency of external force, respectively; and ε is the parameter of coupling of the rotators.

Let us provide examples of physical systems, whose dynamics is modeled by Eq. (5.12).

Example 1 Figure 5.1 shows a chain of Froude pendulums [16]. Sleeves of identical pendulums are mounted on the shaft, which is driven into rotational motion with a constant speed of rotation $\Omega = \omega$, or into rotational-oscillatory motion with frequency $\Omega = \omega + B \sin \Omega_0 t$. There is viscous friction between all sleeves and the shaft as well as between all contacting couplings.

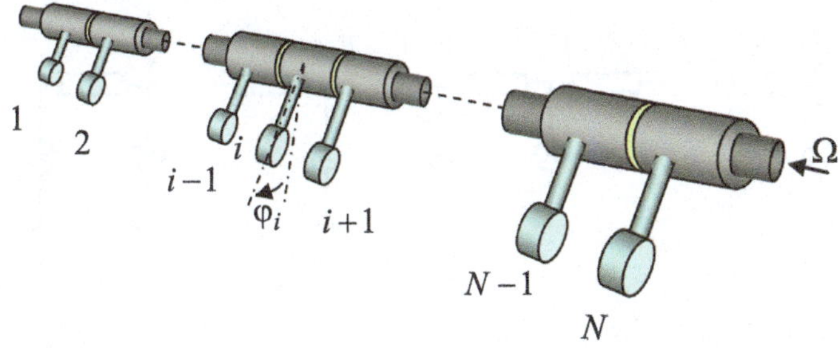

Fig. 5.1 A chain of froude pendulums

In physical variables and parameters, the equations of motion of the chain have
the form:

$$ml^2\ddot{\varphi}_i + d\dot{\varphi}_i + mgl\sin\varphi_i = R(\omega + B\sin\psi - \dot{\varphi}_i)$$
$$+ \lambda(\dot{\varphi}_{i-1} - \dot{\varphi}_i) + \lambda(\dot{\varphi}_{i+1} - \dot{\varphi}_i),$$
$$\dot{\psi} = \Omega_0,$$
$$i = \overline{1, N}.$$

Here, m, l are the reduced mass and length of the pendulum, respectively; d is
the constant coefficient of viscous damping of the medium; $R(\omega + B\sin\psi - \dot{\varphi}_i)$
is the moment of force of viscous friction acting from the shaft side; $\lambda(\dot{\varphi}_{i-1} - \dot{\varphi}_i)$,
$\lambda(\dot{\varphi}_{i+1} - \dot{\varphi}_i)$ are the moments of the friction forces from the neighboring pendulums;
and R, λ are the constant coefficients.

Dividing the equations by mgl and introducing dimensionless time $\frac{mgl}{d+R}t = \tau$, we
obtain Eq. (5.12). The relation between dimensional and dimensionless parameters
is defined by formulae

$$I = \frac{l}{g}\left(\frac{mgl}{d+R}\right)^2, \gamma = \frac{R\omega}{mgl}, \varepsilon = \frac{\lambda}{d+R}, \omega_0 = \Omega_0\frac{d+R}{mgl}, A = \frac{RB}{mgl}.$$

Example 2 Let us remind the reader that a dynamical state of a superconductive
junction is governed by two variables: φ, representing the difference of phases of
quantum-mechanical functions of order, and $\dot{\varphi} = V$, the potential difference of superconductors (the Josephson ratio in dimensionless time). In a resistive model written in
dimensionless variables and parameters, a separate non-autonomous superconductive
junction is described by a pendulum equation (see Sect. 1.1).

Figure 5.2 shows a chain of dissipative-coupled superconductive junctions with
uniform injection of a constant current in an external microwave field.

Equations for electric currents in nodes A_i and the Josephson relation define the
dynamical system of the chain:

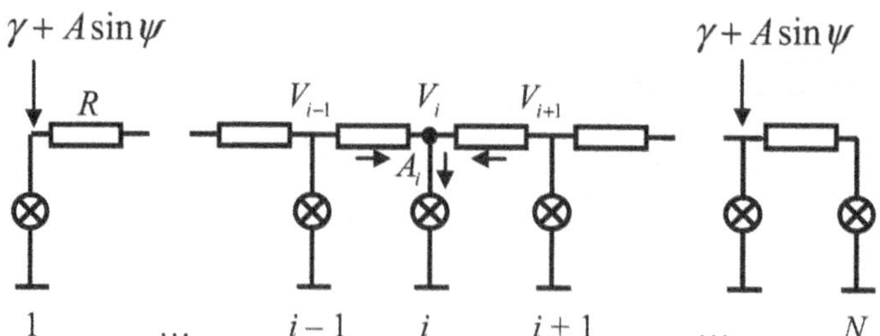

Fig. 5.2 A chain of superconductive junctions

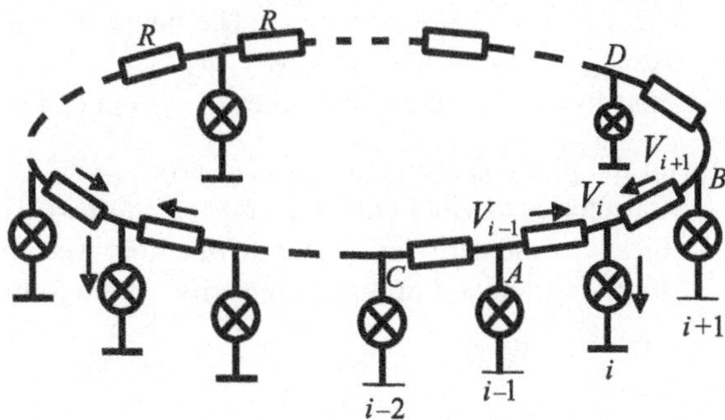

Fig. 5.3 A ring of dissipative-coupled Josephson junctions

$$c\ddot{\varphi}_i + \dot{\varphi}_i + \sin\varphi_i = \gamma + \frac{1}{r}(\dot{\varphi}_{i-1} - \dot{\varphi}_i) + \frac{1}{r}(\dot{\varphi}_{i+1} - \dot{\varphi}_i) + A\sin\psi,$$

$$\dot{\psi} = \omega_0,$$

$$i = \overline{1, N}$$

with boundary conditions $\varphi_0 \equiv \varphi_1$, $\varphi_N \equiv \varphi_{N+1}$. Here, the dimensionless parameters are: c is the capacitance of the junction; γ is the constant electric current of the external power supply; A, ω_0 are the amplitude and frequency of external microwave field, respectively; and r is the resistance of the coupling. Assuming $1/r = \varepsilon$ (dimensionless conductivity), we obtain Eq. (5.12).

Example 3 Figure 5.3 shows a block-scheme of a ring of dissipative-coupled superconductive junctions. By writing the physical equations for each of them, we obtain dynamical system (5.13) with periodic conditions $\varphi_0 \equiv \varphi_N$, $\varphi_{N+1} \equiv \varphi_1$. The count of the sequential number of junctions starts from an arbitrary element.

From a comparison of equations of physical systems outlined in Examples 5.1 and 5.2, we obtain the following electromechanical analogy of the variables: V (voltage), $\sim \dot{\varphi} = \omega$ (frequency of rotation), $1/R$ (conductivity of the coupling), $\sim \varepsilon$ (coefficient of viscous damping); J (current); and $\sim M$ (moment). The expression for the electric current of the coupling (Ohm's law): $(V_{i+1} - V_i)/R = J_{i+1}$ that flows from junction $(i + 1)$ to junction i is equivalent to the expression for the moment of force of viscous friction acting from pendulum $(i + 1)$ on pendulum i. Condition $\langle\dot{\varphi}_i\rangle_t = 0$ corresponds to the equilibrium and oscillatory motions of pendulum i from the chain shown in Fig. 5.1, as well as to the superconductive state of the corresponding junction. Condition $\langle\dot{\varphi}_i\rangle_t \neq 0$ corresponds to the rotational movements of the pendulum and the resistive state, or to the generation mode of the junction.

1^0. *An autonomous chain and a ring.* Assuming $A = 0$ in (5.12) and transforming time $\sqrt{I^{-1}}\tau = \tau_n$, we obtain a system of the form

$$\ddot{\varphi}_i + \lambda\dot{\varphi}_i + \sin\varphi_i = \gamma + \sigma(\dot{\varphi}_{i-1} - 2\dot{\varphi}_i + \dot{\varphi}_{i+1})$$

$$i = \overline{1, N} \tag{5.13}$$

Where $\lambda = \sqrt{I^{-1}}$, $\sigma = \varepsilon\sqrt{I^{-1}}$. The boundary conditions have the form: $\varphi_0 = \varphi_1$, $\varphi_N = \varphi_{N+1}$ for a chain, and $\varphi_0 = \varphi_N$, $\varphi_{N+1} = \varphi_1$ for a ring.

Let us list some general properties of system (5.13).

(1) In the phase space of the system $G(\varphi, \dot\varphi) = T^N \times R^N$, there are no closed trajectories of vibrational type (O-trajectories). This property is established by means of the periodic Lyapunov function: we consider the function (quadratic form plus integral of the nonlinearity) of the form

$$V = \sum_{i=1}^{N} \left(\frac{1}{2}\dot\varphi_i^2 + \int_{\varphi_{0i}}^{\varphi_i} (\sin\varphi_i - \gamma)d\varphi_i \right),$$

Whose derivative taken along trajectories of Eq. (5.13): $\dot V = -\sum_{i=1}^{N} \left(\lambda\dot\varphi_i^2 + \sigma(\dot\varphi_i - \dot\varphi_{i+1})^2 \right) \leq 0$. The negative sign of the derivative in the whole phase space $G(\varphi, \dot\varphi)$ proves the property.

Thus, the possible attractors of system (5.13) can be: equilibrium states, rotational-oscillatory and rotational limit cycles, tori, and, perhaps, chaotic attractors with a rotating phase.

(2) In the phase space of the system, there exists the "main" integral manifold $M_0 = \{\varphi_1 = \varphi_2 = \ldots = \varphi_N, \dot\varphi_1 = \dot\varphi_2 = \ldots = \dot\varphi_N\}$ representing cylinder $(\varphi, \dot\varphi)$, filled by phase trajectories of the equation of a single rotator, i.e. a pendulum equation (see Sect. 5.1):

$$\ddot\varphi + \lambda\dot\varphi + \sin\varphi = \gamma.$$

Properties of the pendulum equation are described in detail in Sect. 1.1.

Let us study stationary, spatially homogeneous dynamical regimes of the chain (synchronization) that correspond to stable limit sets of phase trajectories lying on the main manifold M_0. Stable equilibrium $O_1(\arcsin\gamma, 0)$ corresponds to the resting Froude's pendula, located at the same angle with respect to the vertical. The limit cycle corresponds to the in-phase rotation of all of the pendula of the chain. Using the example of this system, let us repeat the proof that these limit sets are stable along directions that are transverse to M_0.

Using a change of the variables $u_i = \varphi_i - \varphi_{i+1}$, $\dot u_i = v_i$, we reduce Eq. (5.13) to manifold M_0. As a result, we obtain:

$$\begin{aligned}
\dot u_i &= v_i, \\
\dot v_i &= -\lambda v_i - (\cos\xi_i)u_i + \sigma(v_{i-1} - 2v_i + v_{i+1}), \\
i &= \overline{1, N-1}.
\end{aligned} \tag{5.14}$$

With the boundary conditions: $v_0 \equiv 0$, $v_N \equiv 0$. During the transition to system (3.16), the Lagrange theorem is used: $\xi_i \in (\varphi_i, \varphi_{i+1})$.

Note that if $\xi_1 = \varphi_1 = \varphi_2 = \xi_2 = \ldots = \varphi_N = \varphi$, where $\varphi(t)$ is an arbitrary solution and $(\varphi, \dot\varphi) \in M_0$, then system (5.14) is a system in variation with respect to a given manifold.

We rewrite system (5.14) as a single equation of the form

$$\dot{\mathbf{U}} = (\mathbf{I}_{N-1} \otimes \mathbf{J}_0(\xi) - \sigma \mathbf{D}_{N-1} \otimes \mathbf{C})\mathbf{U}, \tag{5.15}$$

Where

$$\mathbf{U} = (u_1, v_1, u_2, v_2 \ldots, u_{N-1}, v_{N-1})^T, \ \mathbf{J}_0 = \begin{pmatrix} 0 & 1 \\ -\cos\xi & -\lambda \end{pmatrix}, \ \mathbf{C} = \begin{pmatrix} 0 & 0 \\ 0 & 1 \end{pmatrix},$$

$$\mathbf{D}_{N-1} = \begin{pmatrix} \begin{matrix} 2 & -1 & 0 & 0 \\ -1 & 2 & -1 & 0 \\ 0 & -1 & 2 & -1 \\ 0 & 0 & -1 & 2 \end{matrix} & & \mathbf{0} \\ & \ddots & \\ \mathbf{0} & & \begin{matrix} 2 & -1 & 0 & 0 \\ -1 & 2 & -1 & 0 \\ 0 & -1 & 2 & -1 \\ 0 & 0 & -1 & 2 \end{matrix} \end{pmatrix}.$$

According to Lemma 5.1, Eq. (5.15) is decomposed into $N-1$ two-dimensional systems with respect to $(x_i, y_i)^T = \mathbf{V}_i$ of the form

$$\begin{aligned} \dot{x}_i &= y_i, \\ \dot{y}_i &= -\lambda\, y_i - (\cos\varphi)x_i - \sigma\lambda_i y_i, \\ i &= \overline{1, N-1}, \end{aligned} \tag{5.16}$$

where $\lambda_i = 4\sin^2 \pi i / 2N$, $i = \overline{1, N-1}$ are the eigenvalues of matrix $\mathbf{D}_{N-1}$.

If $\sigma = 0$, then each of these systems is a stable system in variation (unperturbed system) with respect to the solution that corresponds to one of the stable limit trajectories of the pendulum equation (stable equilibrium, stable limit cycle). On the other hand, if $\sigma \neq 0$, then for any fixed $t = t_0$ there is a right rotation of the vector field of each of systems (5.16) on the trajectories of the unperturbed system. In other words, since all trajectories of the unperturbed system (in the extended phase space) enter inside some cylinder $Z\{t > t_0,\ x^2 + y^2 \leq R^2(t_0)\}$, $\lim\limits_{t_0 \to \infty} R(t_0) = 0$, and stay there, then the same holds for trajectories of the perturbed system. Thus, the stability of each system from (5.16) follows from the stability of the unperturbed system. Consequently, the solution $\mathbf{V} = 0$, and, therefore, $\mathbf{U} = 0$ is also stable.

Physically, this fact is quite understandable: an additional dissipation is introduced into the stable system, which, naturally, only improves its stability.

Thus, the spatially uniform dynamical regime of the chain (synchronization) is stable for any value of diffusive coupling, including arbitrarily small ones, which is confirmed by the result obtained in the previous section.

Let us turn to the bifurcation diagram (see Fig. 1.1). For the parameters from domain (1), and for the initial conditions given in the general case in a small neighborhood of the manifold, the spatially uniform state of the chain is the equilibrium state: all of the pendula of the chain in the course of time come to equilibrium; that is, they hang under one angle. In parameter domain (2), depending on the initial conditions, equilibrium for a spatially homogeneous regime of synchronous rotations of pendula of the chain can be realized. Finally, in parameter domain (3), for any initial conditions given in a neighborhood of the manifold, a spatially homogeneous regime of in-phase rotations is realized.

The same conclusions as those made for a chain are valid for the ring of rotators

2^0. *Non-autonomous chain and a ring*. Let us study dynamical properties of non-autonomous system (5.12) using a method of the averaging in the zone of main resonance.

Following the algorithms for transformation of systems of coupled rotators, using the a change of the variables of the form

$$\dot{\varphi}_i = \omega_0 + \mu F_i(\varphi_i, \psi, \xi), \ F_i = \frac{\cos \varphi_i}{\omega_0} - \frac{A \cos \psi}{\omega_0} + \xi_i, \ \eta_i = \varphi_i - \psi$$

we obtain a system equivalent to (5.12) in a standard form:

$$\dot{\xi}_i = \mu \left(-F_i \frac{\partial F_i}{\partial \varphi_i} - F_i + \sigma(F_{i-1} - 2F_i + F_{i+1}) + \Delta \right),$$

$$\dot{\eta}_i = \mu F_i,$$

$$\dot{\psi} = \omega_0, \tag{5.17}$$

Where $\mu = I^{-1}$, $\gamma - \omega_0 = \mu \Delta$ is the frequency mistuning (zone of the main resonance), and $\eta_i = \varphi_i - \psi$ are the phase mistunings.

Further we assume that the parameters of an individual rotator are selected from region (2) or (3) shown in Fig. 1.1, for which, in the phase space, there is a rotational limit cycle. We assume that $\mu \ll 1$.

Having averaged system (5.17) over a fast –spinning phase and transforming time $\mu \tau = \tau_n$, we obtained a system of the form

$$\dot{\eta}_i = \xi_i,$$

$$\dot{\xi}_i = -\xi_i - \frac{A}{2\omega_0^2} \sin \eta_i + \Delta + \sigma(\xi_{i-1} - 2\xi_i + \xi_{i+1}) \tag{5.18}$$

with boundary conditions $\xi_0 = \xi_1$, $\xi_N = \xi_{N+1}$.

Then, by using a change of time of the form $\sqrt{\frac{A}{2\omega_0^2}}\,\tau_n = \tau_{nn}$, system (5.18) is reduced to equations

$$\ddot{\eta}_i + \lambda_1^r \dot{\eta}_i + \sin \eta_i = \gamma_1^r + \sigma_0(\dot{\eta}_{i-1} - 2_i\dot{\eta} + \dot{\eta}_{i+1}),$$
$$i = \overline{1, N} \tag{5.19}$$

with boundary conditions $\eta_0 \equiv \eta_1$, $\eta_N \equiv \eta_{N+1}$. Here $\lambda_1^r = \sqrt{\frac{2\omega_0^2}{A}}$, $\gamma_1^r = \frac{2\omega_0^2 \Delta}{A}$, $\sigma_0 = \sqrt{\frac{2\omega_0^2}{A}}\sigma$.

As one can see, the averaged system (5.19) has the same form as the system of equations of the autonomous chain. This means that the character of the spatially homogeneous dynamic regime of the chain is determined by the dynamics of the "averaged" rotator:

$$\ddot{\eta} + \lambda_1^r \dot{\eta} + \sin \eta = \gamma_1^r.$$

The bifurcation diagram of this equation on the parameter plane $\left(\gamma_1^r, \lambda_1^r\right)$ up to the designation of the axes coincides with the Fig. 1.1. To interpret it onto the original non-autonomous system (5.17), we only need to make one correction: to replace the Tricomi curve by a narrow strip that corresponds to the chaotic torus-attractors of system (5.17) (see Sect. 1.2). Let us recall that from the principle of the averaging it follows that if L is some limit set of trajectories of the averaged autonomous system, then $L \times S^1$ is a corresponding limit set of the non-autonomous system including conditions of stability of L.

Thus, we have everything that is necessary to describe the dynamics of the chain in the parameter domain under consideration.

For parameters from domain (1), in the chain of non-autonomous rotators (5.12), a spatially uniform regime of synchronous rotations (mutually forced synchronization) will be observed, which is realized for any set of initial conditions of the system. For domain (3), the regime of quasi-periodic rotations of pendula takes place when there is mutual synchronization, but there is no synchronization with an external source (regime of beatings). For domain (2), one or another regime can be realized depending on the initial conditions. According to the property of the autonomous chain proven earlier, the above dynamical regimes are stable for any value of the coupling parameter, including arbitrarily small ones. For parameters near the Tricomi curve, the regime of synchronous and, at the same time, chaotic rotations of pendula (chaotic synchronization) will be observed [17]. In contrast to previous cases, the stability of this dynamical regime is realized for values of the coupling parameter that exceed the threshold value, which depends both on the Lyapunov exponent of attractors and on the number of rotators in the chain (or ring) (see Sect. 5.1).

5.3 Regular and Chaotic Synchronization in a Homogeneous Chain of Dynamical Systems of "Rotator-Oscillator" Type

Dynamics of the *"rotator-oscillator"* system have been studied in Sect. 3.2. In particular, we have established the existence of Lorenz and Feigenbaum chaotic attractors, as well as alternation of chaos in the phase space of this system. Below, we will consider dynamics of a chain of diffusively coupled systems of this type [18, 19], i.e. synchronization of rotary motions of rotators.

A physical example. Figure 5.4 shows a chain of superconductive junctions, each of which is loaded by an oscillatory circuit. We assume that all junctions, as well as all similar elements of the electric schemes, are identical.

Using Kirchhoff's laws (for the nodes of the scheme shown in Fig. 5.4) and the Josephson ratio, and by carrying out the procedure of making physical variables dimensionless, we obtain a dynamical system governing the chain of the form:

$$c\ddot{\varphi}_k + \dot{\varphi}_k + \sin\varphi_k = \gamma + \varepsilon(\dot{\varphi}_{k-1} - 2\dot{\varphi}_k + \dot{\varphi}_{k+1}) + J_k,$$
$$\ddot{J}_k + \delta\dot{J}_k + \omega_0^2 J_k = b\ddot{\varphi}_k + d\dot{\varphi}_k,$$
$$k = \overline{1, N} \tag{5.20}$$

with boundary conditions $\varphi_0 \equiv \varphi_1$, $\varphi_N \equiv \varphi_{N+1}$.

The relation of parameters of system (5.20) with dimensionless parameters of the above scheme is as follows:

$$\delta = \frac{c_0 r_0 r_1 + l_0}{c_0 l_0 r_1}, \quad \omega_0^2 = \frac{1}{c_0 l_0}\left(1 + \frac{r_0}{r_1}\right), \quad b = -\frac{1}{l_0}, \quad d = -\frac{1}{c_0 l_0 r_1}, \quad \varepsilon = \frac{1}{r}.$$

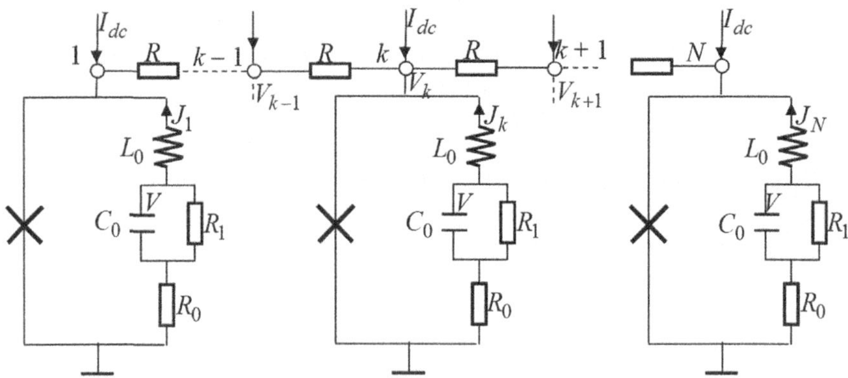

Fig. 5.4 A chain with superconductive junctions

The study of synchronization will be carried out using the technique of averaging that has been discussed in earlier chapters. Consider parameter domain

$$D_\mu = \left\{ c^{-1} = \mu \ll 1, \; \delta = \mu h, \; \gamma + \frac{d}{\omega_0} - \omega_0 = \mu \Delta \right\},$$

that corresponds to the zone of resonance of the rotators and oscillators.

It is not difficult to prove that Eq. (5.20) are equivalent to a system of the form:

$$c\ddot{\varphi}_k + \dot{\varphi}_k + \sin\varphi_k = \gamma + J_k + \varepsilon(\dot{\varphi}_{k-1} - 2\dot{\varphi}_k + \dot{\varphi}_{k+1}),$$

$$\dot{J}_k = W_k + b\dot{\varphi}_k - \mu h J_k,$$

$$\dot{W}_k = -\omega_0^2 J_k + d\dot{\varphi}_k. \tag{5.21}$$

By applying the changes of the variables of the form

$$J_k = \theta_k \sin\varphi_k + \eta_k \cos\varphi_k + \frac{d}{\omega_0},$$

$$W_k = (\theta_k \cos\varphi_k - \eta_k \sin\varphi_k)\omega_0 - b\omega_0,$$

$$\dot{\varphi}_k = \omega_0 + \mu\, \Phi_k(\theta_k, \eta_k, \varphi_k, \xi_k),$$

$$\eta_k = \varphi_k - \varphi_1,$$

where $\Phi_k = \frac{1-\theta_k}{\omega_0}\cos\varphi_k + \frac{\eta_k}{\omega_0}\sin\varphi_k + \xi_k$, we reduce Eqs. (5.21) to an equivalent system in a standard form:

$$\dot{\theta}_k = \mu\Theta_k(\theta_k, \eta_k, \varphi_k, \xi_k),$$

$$\dot{\eta}_k = \mu T_k(\theta_k, \eta_k, \varphi_k, \xi_k),$$

$$\dot{\xi}_k = \mu\Xi_k(\theta_k, \eta_k, \varphi_k, \xi_{k-1}, \xi_k, \xi_{k+1}),$$

$$\dot{\psi}_k = \mu(\Phi_k - \Phi_1),$$

$$\dot{\varphi}_1 = \omega_0 + \mu\Phi_1(\theta_1, \eta_1, \varphi_1, \xi_1) \tag{5.22}$$

The functions on the right-hand sides have the form:

$$\Theta_k = \left(\eta_k + b\sin\varphi_k + \frac{d}{\omega_0}\cos\varphi_k \right)\Phi_k$$

$$- h(\theta_k \sin\varphi_k + \eta_k \cos\varphi_k)\sin\varphi_k + \frac{d}{\omega_0}\sin\varphi_k,$$

$$T_k = \left(b\cos\varphi_k - \theta_k - \frac{d}{\omega_0}\sin\varphi_k \right)\Phi_k$$

$$- h(\theta_k \sin\varphi_k + \eta_k \cos\varphi_k)\cos\varphi_k + \frac{d}{\omega_0}\cos\varphi_k,$$

$$\Xi_k = \Delta - \left(\frac{\partial \Phi_k}{\partial \theta_k}\Theta_k + \frac{\partial \Phi_k}{\partial \eta_k}T_k + \frac{\partial \Phi_k}{\partial \varphi_k}\Phi_k + \Phi_k\right) + \varepsilon(\Phi_{k-1} - 2\Phi_k + \Phi_{k+1}).$$

The averaged system for system (5.22) has the form:

$$\begin{aligned}
\dot{\xi}_k &= \mu(-b_1\xi_k + b_2\theta_k + b_3\eta_k + \Delta + \beta(\xi_{k-1} - 2\xi_k + \xi_{k+1})),\\
\dot{\theta}_k &= \mu(-b_4\theta_k + b_5\eta_k + \eta_k\xi_k + b_6),\\
\dot{\eta}_k &= \mu(-b_4\eta_k - b_5\theta_k - \theta_k\xi_k + b_7),\\
\dot{\psi}_k &= \mu(\xi_k - \xi_1),\\
\dot{\varphi}_1 &= \omega_0 + \mu\xi_1
\end{aligned} \tag{5.23}$$

with boundary conditions: $\xi_0 \equiv \xi_1$, $\xi_N \equiv \xi_{N+1}$ and parameters:

$$b_1 = 1 - \frac{d}{\omega_0^2},\, b_2 = 0,\, b_3 = \frac{1}{2\omega_0^2},\, b_4 = \frac{1}{2}\left(h + \frac{d}{\omega_0^2}\right),\, b_5 = \frac{b}{2\omega_0},\, b_6 = \frac{d}{2\omega_0^2},\, b_7 = \frac{b}{2\omega_0}.$$

A change of the variables of the form

$$\begin{aligned}
&x_k = b_4^{-1}(\xi_k + b_5),\; y_k = (b_3\eta_k + b_2\theta_k)/b_1 b_4 - \Lambda,\\
&z_k = (b_3\theta_k - b_2\eta_k)/b_1 b_4 + R,\; \mu b_4\tau = \tau_n
\end{aligned}$$

Reduces the first $3N$ equations of the averaged system (5.23) to a chain of diffusive-coupled systems of the Lorenz type:

$$\begin{aligned}
\dot{x}_k &= -\sigma(x_k - y_k) + \rho + \varepsilon(x_{k-1} - 2x_k + x_{k+1}),\\
\dot{y}_k &= -y_k + Rx_k - x_k z_k,\\
\dot{z}_k &= -z_k + x_k y_k + \Lambda x_k,\\
k &= \overline{1, N}
\end{aligned} \tag{5.24}$$

with boundary conditions $x_0 \equiv x_1$, $x_N \equiv x_{N+1}$. Parameters in (5.24) are expressed as follows:

$$\sigma = b_1/b_4,\, R = \frac{b_2 b_7 - b_3 b_6}{b_1 b_4^2},\, \Lambda = \frac{b_3 b_7 + b_2 b_6}{b_1 b_4^2},$$

$$\rho = \left(\Delta + (b_1 b_4 b_5 + b_3 b_7 + b_2 b_6)b_4^{-1}\right)b_4^{-2},\, \varepsilon = \frac{\beta}{b_4}.$$

In what follows, we will investigate isochronous synchronization in the chain (5.24) to interpret the results for the original system (5.20).

Dynamical properties of the partial system in (5.24) ($\varepsilon = 0$) have been studied in Sect. 3.2. In particular, by using the following quadratic form

$$V_k = \frac{1}{2}\left(x_k^2 + (y_k + \Lambda)^2 + (z_k - \sigma - R)^2\right)$$

we have established its sphere of dissipation. The derivative of this quadratic form, taken along the vector field of the partial system, has the form

$$\dot{V}_k = -\sigma x_k^2 - (\rho - \sigma\Lambda)x_k - y_k^2 - \Lambda y_k - z_k^2 + (\sigma + R)z_k.$$

From the right-hand side of this equation it follows that there exists such a sphere $U_k : x_k^2 + (y_k + \Lambda)^2 + (z_k - \sigma - R)^2 \le L^2$, outside of which the derivative of the quadratic form $\dot{V}_k < 0$, where L is a constant that depends on parameters σ, R, Λ and ρ. Using this property of the partial system, let us formulate the dissipative property of the coupled system (5.24) in the form of a lemma.

Lemma 5.2 *System (5.24) is dissipative into sphere* $U = \bigcup\limits_{k=1}^{N} U_k$.

Proof Consider the quadratic form $V = \sum\limits_{k=1}^{N} V_k$. Its derivative taken along the vector field of system (5.24) has the form and estimate

$$\dot{V} = \sum_{k=1}^{N} \dot{V}_k - \varepsilon \sum_{k=1}^{N} (x_{k-1} - x_k)^2 \le \sum_{k=1}^{N} \dot{V}_k.$$

Since each of the terms of the last sum is negative outside the sphere U_k, then $\dot{V} \le 0$ outside the sphere of dissipation $U = \bigcup\limits_{k=1}^{N} U_k$ of system (5.24).

This lemma allows us to formulate sufficient conditions for the global asymptotic stability of synchronization in system (5.24) in a form of the following theorem.

Theorem 5.2 *If* $\varepsilon > \alpha/2$*, where*

$$\alpha = \sup_{\forall (x_k, y_k, z_k) \in V_k} \left(\frac{(\sigma + R - z_k)^2 + (y_k + \Lambda)^2}{4} - \sigma \right),$$

Then integral manifold $M_0 = (x_k = x_{k+1},\ y_k = y_{k+1},\ z_k = z_{k+1})$, $k = \overline{1, N}$ *in chain (5.24) with the number of oscillators* $N \le N_0 = [\pi/\arccos \nu]$, *where* $\nu = (2\varepsilon - \alpha)/2\varepsilon$, *is globally asymptotically stable.*

Proof Let us write system (5.24) with respect to variables $u_k = x_k - x_{k+1}$, $v_k = y_k - y_{k+1}$, $w_k = z_k - z_{k+1}$ that represent transversals to manifold M_0:

$$\dot{u}_k = -\sigma(u_k - v_k) + \varepsilon(u_{k-1} - 2u_k + u_{k+1}),$$

$$\dot{v}_k = -v_k + (R - z_k)u_k - x_k w_k + u_k w_k,$$
$$\dot{w}_k = -w_k + (y_k + \Lambda)u_k + x_k v_k - u_k v_k,$$
$$k = \overline{1, N-1}, \ u_0 \equiv 0. \tag{5.25}$$

Consider the Lyapunov function $V = \frac{1}{2}\sum_{k=1}^{N-1} V_k$, $V_k = u_k^2 + v_k^2 + w_k^2$. Its derivative taken along the vector field of system (5.25) has the form

$$\dot{V} = -\sum_{k=1}^{N-1}\left((\sigma + \alpha)u_k^2 + v_k^2 + w_k^2 + (\sigma + R - z_k)u_k v_k + (y_k + \Lambda)u_k w_k\right) - \varepsilon\Phi,$$

$$\Phi = \sum_{k=1}^{N-1}\left(2\nu u_k^2 - 2u_k u_{k+1}\right), \tag{5.26}$$

where $\nu = 1 - \alpha/2\varepsilon$, and α is a parameter that is to be chosen.

According to the Lemma, all phase trajectories of (5.24) reach inside sphere $U = \bigcup_{k=1}^{N} U_k$ and stay there forever, i.e. for $t \to \infty$. Let us select the value of parameter α as follows: $\alpha = \sup_{\forall(x_k, y_k, z_k) \in V_k}\left(\frac{(\sigma + R - z_k)^2 + (y_k + \Lambda)^2}{4} - \sigma\right)$. In this case, each of the quadratic forms under the first sum sign is positive. One just has to determine the conditions for the positivity of quadratic form $\Phi = \sum_{k=1}^{N-1}\left(2\nu U_k^2 - 2U_k U_{k+1}\right)$ in Eq. (5.26).

It can be proven that the values of the principal minors Δ_k of the matrix of quadratic form Φ

$$\begin{pmatrix}
\begin{matrix} 2\nu & -1 & 0 & 0 \\ -1 & 2\nu & -1 & 0 \\ 0 & -1 & 2\nu & -1 \\ 0 & 0 & -1 & 2\nu \end{matrix} & 0 & \mathbf{0} \\
 & & \\
\mathbf{0} & \ddots & 0 \\
 & & \begin{matrix} 2\nu & -1 & 0 & 0 \\ -1 & 2\nu & -1 & 0 \\ 0 & -1 & 2\nu & -1 \\ 0 & 0 & -1 & 2\nu \end{matrix} \\
\mathbf{0} & 0 &
\end{pmatrix}$$

are the solutions of a recurrent equation of the form

$$\Delta_k = 2\nu\Delta_{k-1} - \Delta_{k-2}, \ \Delta_{-1} = 0, \ \Delta_0 = 1, \ k = \overline{1, N-1}.$$

Since $\varepsilon > \alpha/2$, then $0 < \nu < 1$, and in this case, the equation has the following solution:

$$\Delta_k = \frac{\sin(k+1)\theta}{\sin\theta}, \text{ where } \theta = \arccos\nu.$$

After certain simple transformations, we obtain for all $N \le N_0 = [\pi/\arccos\nu]$ values $\Delta_k > 0$, $k = \overline{1, N-1}$. Thus, for the conditions of the aforementioned theorem, quadratic form Φ is positive in the whole phase space, and, therefore, a trivial solution of (5.25) that corresponds to isochronous synchronization in systems (5.24) and (5.20) is globally asymptotically stable.

Thus, for the parameters of Eq. (5.20) determined by the conditions of the theorem, there will be a regime of synchronization of rotations of rotators in the chain that will be realized under any set of initial conditions of the system. In this case, the character of the synchronous rotations will be completely determined by the parameters and the dynamic regimes of the single *"rotator-oscillator"* system studied in Sect. 3.2. By changing a parameter of the system (for example, R), we can successively observe regimes of periodic, quasiperiodic, and chaotic mutual synchronization of rotations of rotators. When observing the dynamics of individual elements of the chain (partial systems) in their phase spaces, a complete identity of the phase portraits is observed (see Sect. 3.2) with synchronous motion of the partial affixes along the trajectories of these phase portraits. Without going into much detail, let us pay attention to the fact that the conditions of local stability of synchronization $\varepsilon > \varepsilon^*/4 \sin^2 \frac{\pi}{2N}$ and conditions of its global stability $\varepsilon > \alpha/4 \sin^2 \frac{\pi}{2N}$ have identical forms.

References

1. Gaponov-Grekhov, A.V., Rabinovich, M.I., Starobinets, I.M.: Lett. ZhTF 39, 561 (1982) (in Russian)
2. Afraimovich, V.S., Rabinovich, M.I., Sbitnev, V.I.: Lett. ZhTF **6**, 338 (1984) (in Russian)
3. Aranson, I.S., Gaponov-Grekhov, A.V., Rabinovich, M.I., Starobinets, I.M.: ZhETF **90**, 1707 (1986) (in Russian)
4. Anishchenko, V.S., Aranson, I.S., Postnov, D.E., Rabinovich, M.I. DAN SSSR **286**(5), 628 (1986) (in Russian)
5. Aranson, I.S., Verichev, N.N.: Dynamics of quasiperiodic excitations in unidirectional chains of generators. Izv. vuzov-Radiofizika **31**(1), 29–40 (1988). (in Russian)
6. Afraimovich, V.S., Nekorkin, V.I., Osipov, G.V., Shalfeev, V.D.: Stability, structures and chaos in nonlinear synchronization networks. In: Gaponov-Grekhov, A.V., Rabinovich, M.I., Gorky (eds.) IPF Academy of Sciences of the USSR (1989) (in Russian)
7. Abarbanel, H.D., Rabinovich, M.I., Selverston, A., Bazhenov, M.V., Huerta, R., Sushchik, M.M., Rubchinskii, L.L.: Synchronisation in neural networks. UFN **166**(4), 363–390 (1996); Phys. Usp. **39**(4), 337–362 (1996)
8. Borisyuk, G.N., Borisyuk, R.M., Kazanovich, Y.B., Ivanitskii, G.R.: Models of neural dynamics in brain information processing—the developments of the decade. Phys. Usp. **45**, 1073–1095 (2002). (in Russian)
9. Carpenter, G.A.: Neural network models for pattern recognition and associative memory. Neural Netw. **2**, 243–257 (1989)
10. Kohonen, T.: Self-Organizing Maps. Springer, Berlin (1997)
11. Tan, Z., Ali, M.K.: Pattern recognition in a neural network with chaos. Phys. Rev. E. **58**, 36–49 (1998)

12. Verichev, N.N.: Stability of structures in nonequilibrium systems. Comput. Mech. Continuous Media **6**(1), 23–33 (2013). (in Russian)
13. Pecora, L.M., Carroll, T.L.: Synchronization in chaotic systems. Phys. Rev. Lett. **64**, 821 (1990)
14. Lancaster, P.: Theory of Matrices. Academic Press (1969)
15. Belykh, V.N., Verichev, N.N.: Spatially homogeneous autowave processes in systems with transport and diffusion. Izv. universities. Ser. Radiophysics. **39**(5), 588–596 (1996). (in Russian)
16. Verichev, N.N., Verichev, S.N., Erofeev, V.I.: Cluster dynamics of a homogeneous chain of dissipatively coupled rotators. App. Math. Mech. **72**(6), 882–897 (2008)
17. Afraimovich, V.S., Verichev, N.N., Rabinovich, M.I.: Stochastic synchronization of oscillation in dissipative systems. Radiophys. Quantum Electron. **29**, 795–803 (1986)
18. Belykh, V.N., Belykh, I.V., Verichev, N.N.: Regular and chaotic spatially-homogeneous oscillations in a chain of coupled superconductive junctions. Izv. Vuzov Radiofizika **15**(7), 912–924 (1997). (in Russian)
19. Belykh, I.V., Verichev, N.N.: Global synchronization and strange attractors in coupled pendulum systems. Bull. NNGU Nonlinear Dyn. Chaos **2**, 93–102 (1997). (in Russian)

Chapter 6
Physics, Existence, Fusion, and Stability of Cluster Structures

A cluster in lattices of dynamical systems is a group of lattice elements with identical dynamic behavior. Dynamical systems that belong to the same cluster can, as both neighboring and not neighboring, be spread over the integer lattice coordinates in a certain way. In this case, if the dynamic state of a lattice is characterized by two or more clusters, then the distribution of their elements along the lattice coordinates forms a cluster structure. In lattices of the correct geometric form (chain, ring, two-dimensional rectangular lattice, etc.), such structures, being visualized by colors, have the form of multicolored (by the number of clusters) "borders" and "parquet floorings" governed by different types of symmetries of these lattices. We note that, in the kaleidoscope of all possible structures, the spatially homogeneous structure that has been studied in the previous Chapter, which is determined by the isochronous synchronization of all lattice oscillators, is the simplest possible structure (all lattice oscillators form a single cluster, i.e. a single-color cluster structure).

Lattices of interest have self-oscillating systems as their elements. For this reason, one can state that, as in the case of a spatially homogeneous structure, the uniformity of cluster dynamics is determined by the synchronization processes. However, this synchronization reveals itself in a rather unusual way. In this context, questions arise regarding the nature of the "unusual" synchronization entities formed in the lattice during a self-organization and determination of non-trivial cluster structures (these are no longer single oscillators). In parallel to that, questions arise about the role of the geometric parameters of lattices, about the influence of various types of lattice symmetries on the number and type of all possible cluster structures, as well as about questions related to the stability of cluster structures. This chapter is devoted to providing answers to these and other questions related to cluster structures. In what follows, we will use illustration material from the dynamics of oscillators (they are more convenient for numerical experiments and demonstration), but all that will be said fully applies to the main subject of interest, i.e. to the dynamics of rotators.

© Springer Nature Switzerland AG 2020

N. Verichev et al., *Chaos, Synchronization and Structures in Dynamics of Systems with Cylindrical Phase Space*, Understanding Complex Systems,
https://doi.org/10.1007/978-3-030-36103-7_6

6.1 Physics of Cluster Structures

We discuss the existing interpretation of cluster structures [1–8] in homogeneous lattices using the example of a chain of oscillators whose equations, for reasons of convenience, we rewrite as follows:

$$\dot{X}_i = \mathbf{F}(\mathbf{X}_i, t) + \varepsilon \mathbf{C}(X_{i-1} - 2X_i + X_{i+1}),$$
$$i = \overline{1, N}, \quad \mathbf{X}_0 \equiv \mathbf{X}_1, \quad \mathbf{X}_N \equiv \mathbf{X}_{N+1}. \tag{6.1}$$

As stated before, system (6.1) has integral manifold $M_0 = \left\{\mathbf{X}_i = \mathbf{X}_{i+1}, i = \overline{1, N}\right\}$. Its existence is obvious, and it has played a role in the study of the synchronization of oscillators (rotators), which generated a spatially uniform state of the lattice. It turns out, however, that in addition to M_0, there also exist other ("cluster") invariant manifolds that have the form [5]:

$$M_{N,n+1} = \{\mathbf{X}_{2n+1} = \mathbf{X}_1, \mathbf{X}_{2n} = \mathbf{X}_2, .., \mathbf{X}_{n+2} = \mathbf{X}_n\}, \text{ if } N = 2n + 1;$$
$$M_{N,n} = \{\mathbf{X}_{2n} = \mathbf{X}_1, \mathbf{X}_{2n-1} = \mathbf{X}_2, \ldots, \mathbf{X}_{n+1} = \mathbf{X}_n\}, \text{ if } N = 2n;$$
$$M_{pq,q} = \left\{\mathbf{X}_i = \mathbf{X}_{i+2qj}, j = 1, 2, .., [(p-1)/2], i = \overline{1, q}\right\},$$
$$M_{pq,q} = \left\{\mathbf{X}_i = \mathbf{X}_{-i+1+2qj}, j = 1, 2, .., [p/2], i = \overline{1, q}\right\}, \text{ if } N = pq,$$

where p, q are integers.

If we assume that some of the manifolds are stable, then its equations also represent an analytic expression for the corresponding cluster structure. In particular, if $N = 6$, then there are two such manifolds:

$$M_{6,3} = \{\mathbf{X}_6 = \mathbf{X}_1, \mathbf{X}_5 = \mathbf{X}_2, \mathbf{X}_4 = \mathbf{X}_3\} \text{ and}$$
$$M_{6,2} = \{\mathbf{X}_1 = \mathbf{X}_4 = \mathbf{X}_5, \mathbf{X}_2 = \mathbf{X}_3 = \mathbf{X}_6\}.$$

Depending on the initial conditions, one of the cluster structures shown schematically in Fig. 6.1 will be realized in the chain. In this figure, the equally colored squares correspond to the oscillators of the chain with identical dynamics. In what follows, similar images will be referred to as schemes of cluster structures.

The dynamics of lattices on manifolds has a name "cluster synchronization" and was studied in papers [3, 6–11].

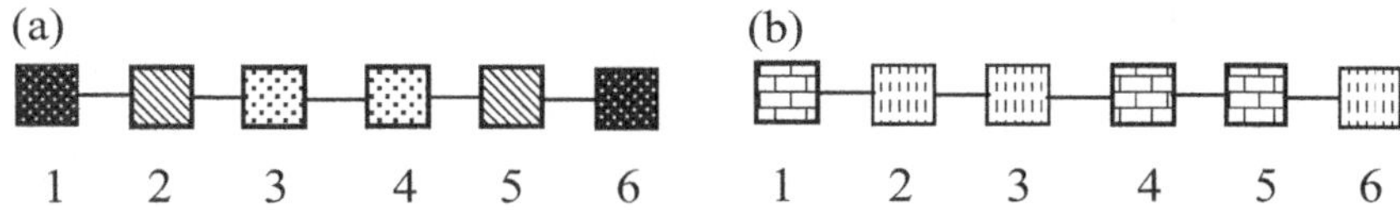

Fig. 6.1 Cluster structures in a chain of $N = 6$ oscillators

Below, we will formulate everything using a simple and clear language of classical synchronization [12–14], without appealing to their cluster manifolds that will be obtained (if necessary) as a consequence.

Subjects of cluster structures. Observations of cluster structures in lattices of different geometric dimensions and shapes lead to a simple conclusion that subjects of cluster synchronization represent "generalized" oscillators or, in other words, group oscillators that self-organize as separate and indivisible objects composed of elementary oscillators. Below, we arrive at this conclusion by induction, considering the dynamics of oscillator systems starting with simple examples.

Consider a system of two identical, coupled oscillators of the form

$$\dot{\mathbf{X}}_1 = \mathbf{F}(\mathbf{X}_1) + \varepsilon\mathbf{C}(\mathbf{X}_2 - \mathbf{X}_1),$$
$$\dot{\mathbf{X}}_2 = \mathbf{F}(\mathbf{X}_2) + \varepsilon\mathbf{C}(\mathbf{X}_1 - \mathbf{X}_2), \tag{6.2}$$

where $\mathbf{C}$ is a matrix that determines the structure of coupling of the elementary oscillators, and ε is a scalar parameter. It is clear that in the usual case all stationary dynamic modes of this system can be divided into two types:

(1) a regime of isochronous regular or chaotic mutual synchronization. This regime corresponds to the solution $\mathbf{X}_1(t) = \mathbf{X}_2(t) = \mathbf{X}(t)$. The form of the synchronous movements of the oscillators is determined by the parameters and type of the attractor (regular, chaotic) of a single oscillator $A(1)$. Important: in this case, coupled system (6.2) decomposes into a pair of synchronized oscillators;

(2) a regime of non-isochronous dynamics that corresponds to attractor $A_s(2)$. This regime can represent either stationary beatings (regular or chaotic), or non-isochronous synchronization. In case when the affix moves over attractor $A_s(2)$, the following is valid: $\mathbf{X}_1(t) \neq \mathbf{X}_2(t)$. We note that, in this case, system (6.2) is a single object that is indivisible in this sense.

To illustrate what has been said, Fig. 6.2 shows the regime of chaotic synchronization and the regime of beatings of chaotic Chua self-generators [15, 16]. A single Chua's oscillator is described by a dynamical system of the form

$$\dot{x} = \alpha(y - h(x)),$$
$$\dot{y} = -y + z - x,$$
$$\dot{z} = -\beta y - \gamma z.$$

Here $h(x) = m_1 x + \frac{m_0 - m_1}{2}(|x + 1| - |x - 1|)$, $m_0 < 0$, $\alpha, \beta, \gamma, m_1 > 0$ are the parameters.

Parameters of the experiment: $(\alpha, \beta, \gamma, m_0, m_1, \varepsilon) = (9.5; 14; 0.1; -1/7; 2/7; 1.0)$, the matrix of couplings of oscillators: $\mathbf{C} = \mathrm{diag}(1, 0, 0)$.

As we see in the first case, the phase portraits are completely identical, and this is additionally confirmed by the diagonal in the phase plane of the same-titled variables (the third one in the upper row). This is the projection of the attractor of the system onto the given plane. In the second case, the projection is different from the diagonal.

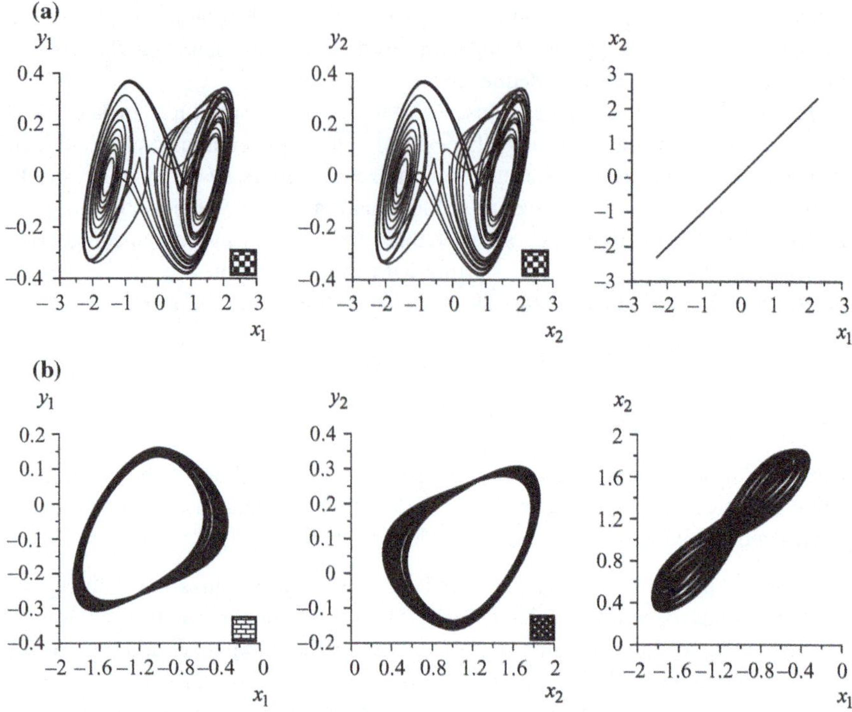

Fig. 6.2 Dynamical regimes of system (6.2): **a** attractor A (1) of the regime of isochronous synchronization; **b** attractor A_s (2), which corresponds to the regime of chaotic beatings in projections onto the coordinate planes of the oscillators

Schematically, the dynamics of system (6.2) in the indicated regimes are shown in Fig. 6.3.

As was previously said, in the regime of synchronization, system (6.2) decomposes into a pair of synchronized oscillators (by the condition of exact synchronism). Physically, this fact corresponds to the removal of the coupling between the corresponding oscillators without violating their dynamic regime. In particular, if the oscillators physically represent radio-frequency self-generators, then this procedure

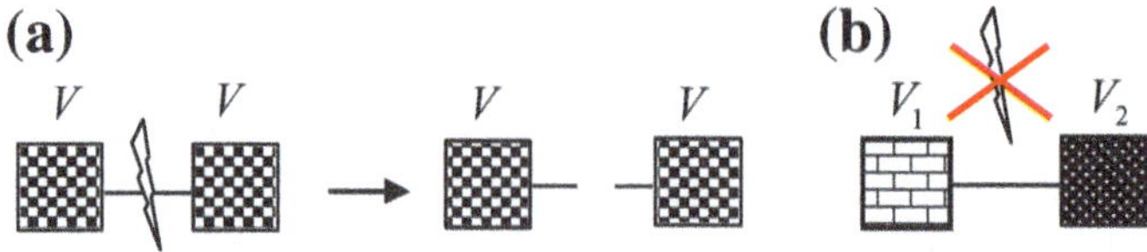

Fig. 6.3 Schematic representation of regimes 1 and 3 of the coupled system. A symbolic knife is shown above the coupling of the oscillators

corresponds to cutting the connections between equipotential points, which represents one of the methods of equivalent transformations of electric circuits. We note that a spatially homogeneous structure is cut into individual oscillators, while the cluster structure (see Fig. 6.1) is cut into identical blocks. In the future, such a procedure of cutting the schemes of cluster structures (couplings between "same-colored" oscillators) will be used as a method for their equivalent transformations, as well as, on the contrary, connection of oscillator blocks through the corresponding points (connection of "same-colored" oscillators) during the fusion of cluster structures.

Let us turn again to system (6.2). Suppose there is attractor $A_s(2)$ in this system and its motions take place along this attractor. In this case, the system represents an indivisible object, i.e. a generalized oscillator, the dynamics of which are described by a vector equation of the form (system (6.2) rewritten in vector form)

$$\dot{\mathbf{X}} = \mathbf{H}(\mathbf{X}), \tag{6.3}$$

where $\mathbf{X} = (\mathbf{X}_1, \mathbf{X}_2)^T$, $\mathbf{H}(\mathbf{X}) = (\mathbf{F}(\mathbf{X}_1), \mathbf{F}(\mathbf{X}_2))^T + \varepsilon \mathbf{B}_s \otimes \mathbf{C}\mathbf{X}$, $\mathbf{B}_s = \begin{pmatrix} -1 & 1 \\ 1 & -1 \end{pmatrix}$.

We will call (6.3) the cluster forming oscillator (or C-oscillator) and denote it as $O_s(2)$. Index s reflects the symmetry of cluster matrix $\mathbf{B}_s$, and 2 is the number of elementary oscillators in the cluster forming oscillator that produces clusters (the exact definition of the cluster forming oscillator will be given below). Schematically, the cluster forming oscillator in the stationary "operational" mode is shown in Fig. 6.3b. Note that the vector coordinates $\mathbf{X}_1$, $\mathbf{X}_2$ play the same role as scalar coordinates of the elementary oscillator.

Suppose now that we have an exact dynamic copy of a given cluster forming oscillator (CFO), which is described by a vector variable $\mathbf{Y} = (\mathbf{Y}_1, \mathbf{Y}_2)^T$ (dynamic processes in the originals and in the copies happen synchronously). The answer to the question of how these two CFOs should be connected in order not to violate their dynamic regimes is contained in the previous reasoning regarding the cutting/connecting of equipotential points that correspond to the same coordinates. In other words, the couplings of a cluster forming oscillator in the synthesized lattice should have the form $(\mathbf{C}^* \otimes \mathbf{C})(\mathbf{X} - \mathbf{Y})$ or $(\mathbf{C}_* \otimes \mathbf{C})(\mathbf{X} - \mathbf{Y})$, where $\mathbf{C}$ is an already known matrix that determines the number of coupled scalar variables of elementary oscillators, $\mathbf{C}^* = \mathrm{diag}(1, 0)$, $\mathbf{C}_* = \mathrm{diag}(0, 1)$ are the matrices that determine the number of coupled oscillators themselves that belong to one and to another cluster forming oscillator (matrices of fusion). The first matrix determines a coupling along the first coordinates $\mathbf{X}_1$, $\mathbf{Y}_1$, while the second one determines a coupling along the second coordinates $\mathbf{X}_2$, $\mathbf{Y}_2$.

Taking the aforesaid matters into account, we consider a system of two coupled cluster forming oscillators of the form

$$\dot{\mathbf{X}} = \mathbf{H}(\mathbf{X}) + \varepsilon(\mathbf{C}_* \otimes \mathbf{C})(\mathbf{Y} - \mathbf{X}),$$
$$\dot{\mathbf{Y}} = \mathbf{H}(\mathbf{Y}) + \varepsilon(\mathbf{C}_* \otimes \mathbf{C})(\mathbf{X} - \mathbf{Y}). \tag{6.4}$$

First, we note that Eqs. (6.2) and (6.4) do not differ in form. Second, if system (6.4) is written coordinate-wise (with respect to elementary oscillators), assuming in this case that $\mathbf{Y}_1 = \mathbf{X}_4$, $\mathbf{Y}_2 = \mathbf{X}_3$, then it represents a chain of 4 elements.

Third, this system, in the case of classical synchronization of a CFO that is expressed by equality $\mathbf{X} = \mathbf{Y}$, defines the cluster structure shown in Fig. 6.1b $((\mathbf{X}_1, \mathbf{X}_2)^T = (\mathbf{Y}_1, \mathbf{Y}_2)^T$ or $\mathbf{X}_1 = \mathbf{X}_4, \mathbf{X}_2 = \mathbf{X}_3)$, for which the oscillators with numbers 5 and 6 are removed. For a complete picture, one should couple an extra CFO to system (6.4).

Suppose that we have one more dynamical copy of cluster forming oscillator $O_s(2)$, described by vector $\mathbf{Z} = (Z_1, Z_2)^T$. By coupling this cluster forming oscillator with the second cluster forming oscillator in system (6.4), we obtain a system of the form

$$
\begin{aligned}
\dot{\mathbf{X}} &= \mathbf{H}(\mathbf{X}) + \varepsilon(\mathbf{C}_* \otimes \mathbf{C})(\mathbf{Y} - \mathbf{X}), \\
\dot{\mathbf{Y}} &= \mathbf{H}(\mathbf{Y}) + \varepsilon(\mathbf{C}_* \otimes \mathbf{C})(\mathbf{X} - \mathbf{Y}) + \varepsilon(\mathbf{C}^* \otimes \mathbf{C})(\mathbf{Z} - \mathbf{Y}), \\
\dot{\mathbf{Z}} &= \mathbf{H}(\mathbf{Z}) + \varepsilon(\mathbf{C}^* \otimes \mathbf{C})(\mathbf{Y} - \mathbf{Z}).
\end{aligned}
\tag{6.5}
$$

The same can be repeated for (6.5) as has been done for (6.4). Then, supposing that $\mathbf{Z}_1 = \mathbf{X}_5$, $\mathbf{Z}_2 = \mathbf{X}_6$, we can see that this system represents a reformatted chain of 6 elements. Secondly, the classical synchronization of this triple cluster forming oscillator generates the cluster structure depicted in Fig. 6.1b.

Consider one more example. Assume we have cluster forming oscillator of the type $O_s(3)$, composed of three elementary oscillators: $\mathbf{X} = (\mathbf{X}_1, \mathbf{X}_2, \mathbf{X}_3)^T$, as well as its copy with vector $\mathbf{Y} = (\mathbf{Y}_1, \mathbf{Y}_2, \mathbf{Y}_3)^T$. Equations for $O_s(3)$ have the same form as (6.3) with $\mathbf{B}_s = \begin{pmatrix} -1 & 1 & 0 \\ 1 & -2 & 1 \\ 0 & 1 & -1 \end{pmatrix}$.

A synthesized pair of such cluster forming oscillators will have equations of the same form as (6.4), where the fusion matrix is replaced by $\mathbf{C}_* = \mathrm{diag}(0, 0, 1)$.

In the regime of synchronization of these cluster forming oscillators, we obtain $(\mathbf{X}_1, \mathbf{X}_2, \mathbf{X}_3)^T = (\mathbf{Y}_1, \mathbf{Y}_2, \mathbf{Y}_3)^T$. In the case of $\mathbf{Y}_3 = \mathbf{X}_4, \mathbf{Y}_2 = \mathbf{X}_5, \mathbf{Y}_1 = \mathbf{X}_6$, we obtain equations $\mathbf{X}_1 = \mathbf{X}_6, \mathbf{X}_2 = \mathbf{X}_5, \mathbf{X}_3 = \mathbf{X}_4$, i.e. we obtain the cluster structure shown in Fig. 6.1a.

To illustrate the aforesaid scenario, Fig. 6.4 shows cluster attractor $A_s(3)$ of cluster forming oscillator $O_s(3)$, obtained in a numerical experiment with a homogeneous chain of three Chua's oscillators for the following parameter set: $(\alpha, \beta, \gamma, m_0, m_1, \varepsilon) = (9.5;\ 14;\ 0.1;\ -1/7; 2/7; 0.37)$.

Figure 6.5 shows a cluster structure in a chain of six Chua's oscillators as a result of classical synchronization of two cluster forming oscillators $O_s(3)$.

As an intermediate conclusion, we state that in a chain of N elements, the number of cluster structures that can be synthesized, based on the $O_s(n)$ type of cluster forming oscillators, is equal to the number of dividers of number N, and, in each specific case, all of these structures are representable.

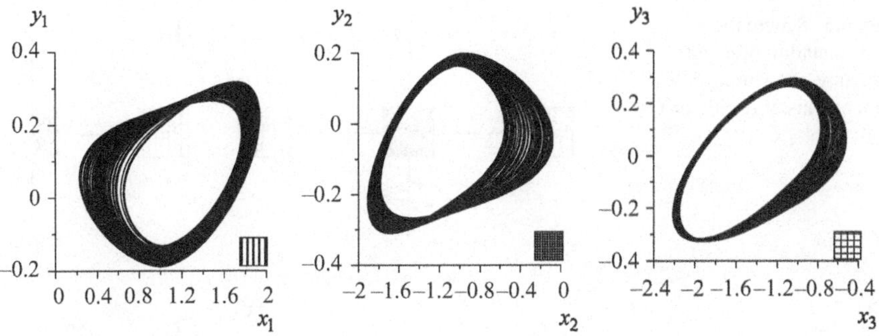

Fig. 6.4 Projections of cluster attractor $A_s(3)$ of cluster forming oscillator $O_s(3)$ onto the coordinate planes

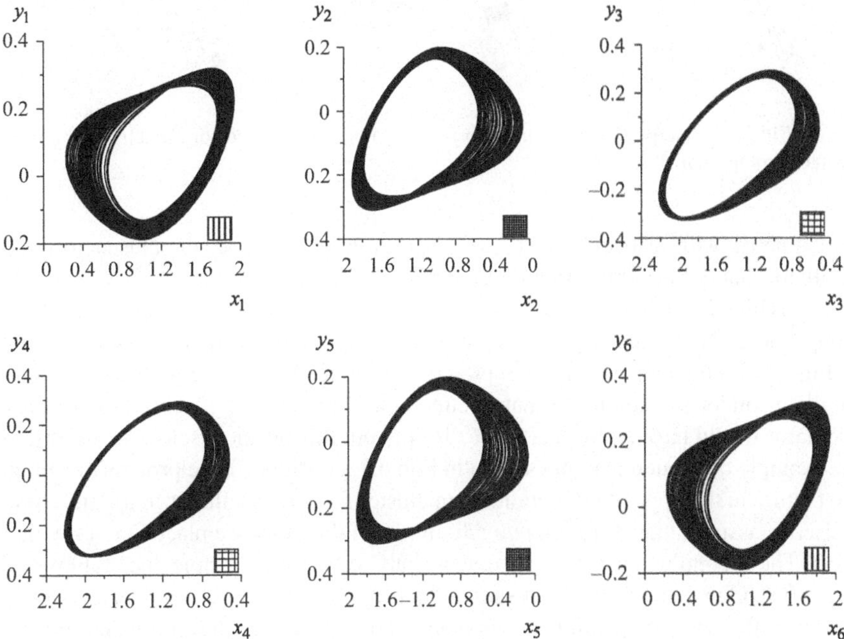

Fig. 6.5 Cluster structure in a homogeneous chain of $N = 6$ Chua's oscillators as a synchronization of two cluster forming oscillators $O_s(3)$

These types of cluster forming oscillators do not exhaust the types of cluster structures in the chains. In particular, if number N is simple, then we can confidently say that the cluster structure in such a chain will belong to another type. Further we shall consider structures in chains with an odd number of oscillators $N = 2k + 1$.

Consider a simple case of a cluster structure in a chain of three elementary oscillators, shown in Fig. 6.6a. It can be seen that this structure contains two clusters and

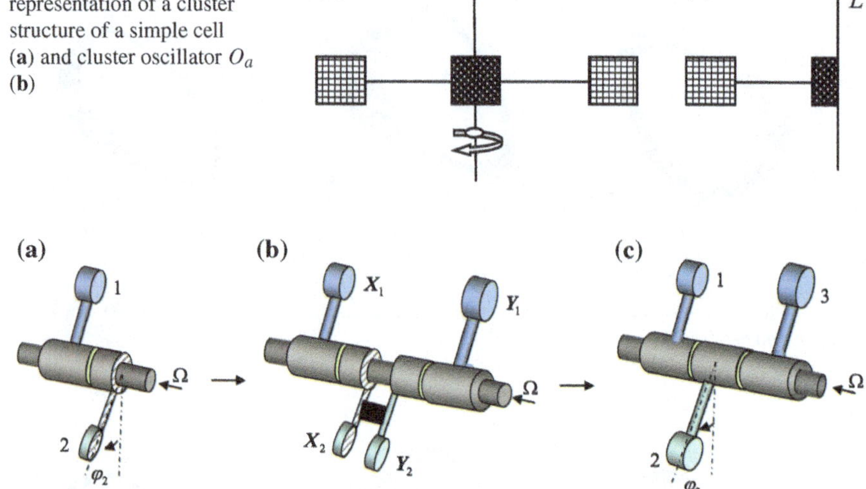

Fig. 6.6 Schematic representation of a cluster structure of a simple cell (**a**) and cluster oscillator O_a (**b**)

Fig. 6.7 Physical interpretation of cluster forming oscillator $O_a(2)$ and simple cell based on this cluster forming oscillator

is symmetric with respect to axis L. The question is: how does the cluster forming oscillator that defines this structure look like?

Let us imagine that this picture is printed on a sheet of paper and this paper is folded along line L. By rolling the sheet, we obtain the schematic representation shown in Fig. 6.6b. One can see that this picture can pretend to depict a cluster forming oscillator under the condition that obtained "half-picture" of a central elementary oscillator would have a physical sense. To find out this physical sense, let us turn to the example of Froude pendula shown in Fig. 6.7, considering the process of rolling the picture in the opposite direction: from cluster forming oscillator to the structure.

Let us assume that there are two adjoining pendula that are placed on a rotating shaft. The second pendulum represents a "half" of the first "whole" pendulum (see Fig. 6.7a). One can imagine that such a "half-pendulum" is obtained as a result of "sawing" the "whole" pendulum in a plane perpendicular to the sleeve and passing through it in the middle. Suppose that there is viscous friction between the sleeves and the shaft. The same friction also exists between the sleeves of pendula. The equations of motion of the system have the form

$$ml^2\ddot{\varphi}_1 + d\dot{\varphi}_1 + mgl\,\sin\,\varphi_1 = R(\Omega - \dot{\varphi}_1) + \lambda(\dot{\varphi}_2 - \dot{\varphi}_1),$$

$$\frac{m}{2}l^2\ddot{\varphi}_2 + \frac{d}{2}\dot{\varphi}_2 + \frac{m}{2}gl\,\sin\,\varphi_2 = \frac{R}{2}(\Omega - \dot{\varphi}_2) + \lambda(\dot{\varphi}_1 - \dot{\varphi}_2).$$

Here $R(\Omega - \dot{\varphi}_1)$ is the moment of force of viscous friction acting on the first pendulum and $\frac{R}{2}(\Omega - \dot{\varphi}_2)$ is the moment of force of viscous friction acting on the second pendulum (they are proportional to the contact area between the sleeve of the pendulum and the shaft); $\lambda(\dot{\varphi}_2 - \dot{\varphi}_1)$ and $\lambda(\dot{\varphi}_1 - \dot{\varphi}_2)$ are the mutual moments of the friction forces of the pendula; and d is the coefficient of the viscous friction of the medium.

After the introduction of dimensionless time $\frac{mgl}{d+R}t = \tau$, we obtain the dynamical system for cluster forming oscillator $O_a(2)$:

$$\dot{\mathbf{X}}_1 = \mathbf{F}(\mathbf{X}_1) + \varepsilon C(-\mathbf{X}_1 + \mathbf{X}_2),$$
$$\dot{\mathbf{X}}_2 = \mathbf{F}(\mathbf{X}_2) + 2\varepsilon C(\mathbf{X}_1 - \mathbf{X}_2), \tag{6.6}$$

where $\mathbf{X}_1 = (\varphi_1, \xi_1)^T$, $\mathbf{X}_2 = (\varphi_2, \xi_2)^T$, $\mathbf{F}(\mathbf{X}_1) = \left(\gamma + I^{-1}\xi_1, -\xi_1 - \sin\varphi_1\right)^T$, $\mathbf{F}(\mathbf{X}_2) = \left(\gamma + I^{-1}\xi_2, -\xi_2 - \sin\varphi_2\right)^T$, $C = \mathrm{diag}(0, 1)$, $I = \frac{l}{g}\left(\frac{mgl}{d+R}\right)^2$, $\gamma = \frac{R\Omega}{mgl}$, $\varepsilon = \frac{\lambda I}{d+R}$.

One can see that, in contrast to cluster forming oscillator $O_s(2)$ (6.2), in which elementary oscillators are coupled by matrix $\mathbf{B}_s = \begin{pmatrix} -1 & 1 \\ 1 & -1 \end{pmatrix}$, and in the case of cluster forming oscillator $O_a(2)$ (6.6), the matrix of couplings is an asymmetric matrix $\mathbf{B}_a = \begin{pmatrix} -1 & 1 \\ 2 & -2 \end{pmatrix}$. Naturally, the difference in the structure of the couplings fundamentally affects the conditions of existence and the character of cluster attractors of these cluster forming oscillators. However, the main difference between $O_s(2)$ and $O_a(2)$ consists of the fact that if the first one is a full-fledged subject of cluster structures, then the role of the second in this sense is limited. Figures 6.7b,c show the process of combining the original $O_a(2)$ with its mirrored copy and the subsequent formation of a clustered and indivisible object of a new type, which we will hereinafter call the simple cell. The process of formation of the cluster structure of a simple cell can also be interpreted as follows: after establishing a rigid coupling between the half-oscillators (note the jumper in Fig. 6.7b), their synchronization occurs immediately (and, in fact, they become a "whole" oscillator). The side oscillators that indirectly interact with each other (through the central oscillator) synchronize their movements.

Let us repeat the aforesaid scenario using the language of equations. We rewrite system (6.6) in the form of one equation:

$$\dot{\mathbf{X}} = \mathbf{H}(\mathbf{X}), \tag{6.7}$$

where $\mathbf{X} = (\mathbf{X}_1, \mathbf{X}_2)^T$, $\mathbf{H}(\mathbf{X}) = (\mathbf{F}(\mathbf{X}_1), \mathbf{F}(\mathbf{X}_2))^T + \varepsilon\mathbf{B}_a \otimes C\mathbf{X}$, $\mathbf{B}_a = \begin{pmatrix} -1 & 1 \\ 2 & -2 \end{pmatrix}$.

Let us suppose that we have an exact copy of this cluster forming oscillator described by vector $\mathbf{Y} = (\mathbf{Y}_1, \mathbf{Y}_2)^T$.

Consider a system of a pair of cluster forming oscillators of the form

$$\dot{\mathbf{X}} = \mathbf{G}(\mathbf{X}) + \varepsilon\mathbf{D} \otimes C(-\mathbf{X} + \mathbf{Y}),$$

$$\dot{\mathbf{Y}} = \mathbf{G}(\mathbf{Y}) + \varepsilon \mathbf{D} \otimes C(\mathbf{X} - \mathbf{Y}). \tag{6.8}$$

Here $\mathbf{X} = (\mathbf{X}_1, \mathbf{X}_2)^T$, $\mathbf{Y} = (\mathbf{Y}_1, \mathbf{Y}_2)^T$, $\mathbf{D} = \begin{pmatrix} 0 & 0 \\ 1 & \lambda \end{pmatrix}$, $\lambda > 0$ is a parameter responsible for the magnitude of couplings of "half-oscillators". If λ has moderate values, then system (6.8) is, in principle, similar to system (6.4), including the synchronization-generated cluster structure corresponding to solution $\mathbf{X} = \mathbf{Y}$ (see Fig. 6.1b without the fifth and sixth elements). However, if $\lambda \to \infty$ (a rigid "jumper" is installed between half-oscillators), then one can show that $\|\mathbf{X}_2 - \mathbf{Y}_2\| = e^{-2\varepsilon\lambda t} \to 0$. That is, as expected, there is a degeneration of the number of degrees of freedom by one elementary oscillator in system (6.8) (the "halves" are glued together). Assuming now that $\mathbf{X}_2 \equiv \mathbf{Y}_2$, $\mathbf{Y}_1 = \mathbf{X}_3$ and by introducing a new vector $\mathbf{X}: \mathbf{X} = (\mathbf{X}_1, \mathbf{X}_2, \mathbf{X}_3)^T$, instead of system (6.8), we obtain a system of the form

$$\dot{\mathbf{X}} = \mathbf{S}(\mathbf{X}), \tag{6.9}$$

where $\mathbf{S}(\mathbf{X}) = (\mathbf{F}(\mathbf{X}_1), \mathbf{F}(\mathbf{X}_2), \mathbf{F}(\mathbf{X}_3))^T + \varepsilon \mathbf{B}_s \otimes C\mathbf{X}, \mathbf{B}_s = \begin{pmatrix} -1 & 1 & 0 \\ 1 & -2 & 1 \\ 0 & 1 & -1 \end{pmatrix}$.

One can see that Eq. (6.9) coincides by its form with the equation of cluster forming oscillator $O_s(3)$. However, its internal content is fundamentally different. First, this equation is considered under the condition of the existence of attractor $A_a(2)$. Second, Eq. (6.6) is considered in a certain part of the phase space: either in the vicinity of attractor $A_a(2)$. (during the study of stability of structures), or on the attractor itself, as long as one deals with a fusion of schemes of the corresponding cluster structures. For these conditions, Eq. (6.9) represents a structured (or potentially structured, see Fig. 6.6a) and indivisible (in a sense) object: a simple cell. Considering the aforesaid matter, a simple cell can be interpreted as a limiting case of a system of two coupled cluster forming oscillators of the type $O_a(n)$. Figure 6.8 shows a "working" attractor of cluster forming oscillator $O_a(2)$, which has been obtained during a numerical study of system (6.6) of two Chua's oscillators. The following parameter set has

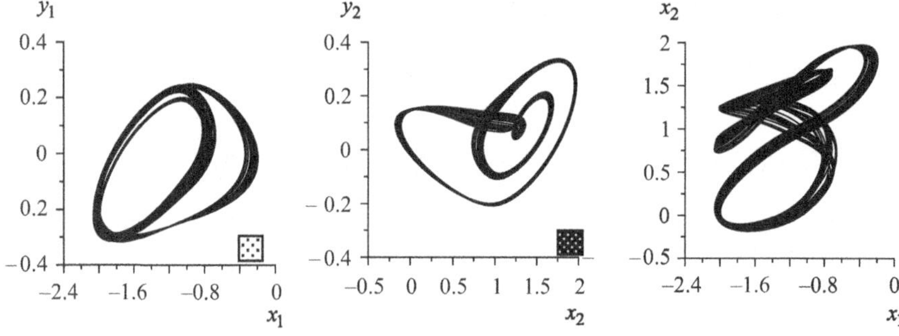

Fig. 6.8 Attractor $A_a(2)$ of cluster forming oscillator $O_a(2)$ in projections into coordinate planes

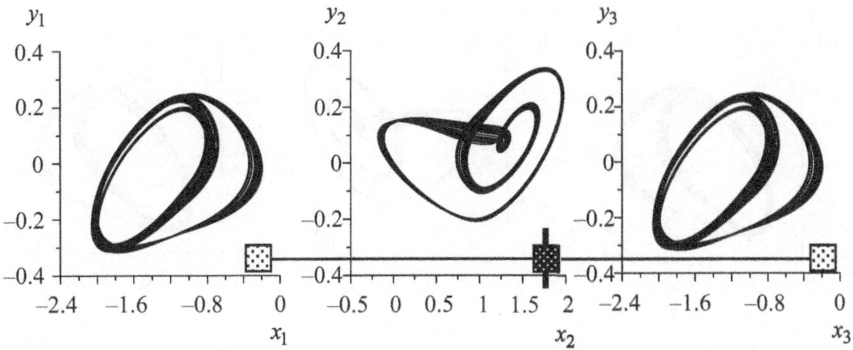

Fig. 6.9 The main dynamical regime: cluster structure of a simple cell based on O_a (2)

been used: $(\alpha, \beta, \gamma, m_0, m_1, \varepsilon) = (9.5; \ 14; \ 0.1; \ -1/7; 2/7; 0.215)$ as well as matrix $C = \mathrm{diag}(1, 0, 0)$.

Figure 6.9 shows an "operational" regime of simple cell (6.9) that represents a dynamical cluster structure based on O_a (2). The following parameter set has been used: $(\alpha, \beta, \gamma, m_0, m_1, \varepsilon) = (9.5; \ 14; \ 0.1; \ -1/7; \ 2/7; \ 0.215)$, $C = \mathrm{diag}(1, 0, 0)$.

Thus, a simple cell, similarly to the cluster forming oscillator of the type O_s (n), also represents a subject in forming cluster structures in lattices. Moreover, because of the symmetry of its own cluster structure, any matrix, C_* or C^*, can be the fusion matrix for one-dimensional lattices (chain, ring) with a cluster structure based on simple cells. Let us illustrate this using the language of equations.

Suppose there is a simple cell (6.9) with a certain cluster structure and its dynamical copy is described by vector $\mathbf{Y} = (\mathbf{Y}_1, \mathbf{Y}_2, \mathbf{Y}_3)^T$. Consider a system of the form

$$\dot{\mathbf{X}} = \mathbf{S}(\mathbf{X}) + \varepsilon(\mathbf{C}_* \otimes \mathbf{C})(\mathbf{Y} - \mathbf{X}),$$

$$\dot{\mathbf{Y}} = \mathbf{S}(\mathbf{Y}) + \varepsilon(\mathbf{C}_* \otimes \mathbf{C})(\mathbf{X} - \mathbf{Y}), \tag{6.10}$$

where $\mathbf{C}_* = \mathrm{diag}(0, 0, 1)$.

From the condition of synchronization in (6.10) and the form of the structure of SC, we obtain $(\mathbf{X}_1, \mathbf{X}_2, \mathbf{X}_3)^T = (\mathbf{Y}_1, \mathbf{Y}_2, \mathbf{Y}_3)^T$ or $\mathbf{X}_1 = \mathbf{X}_3 = \mathbf{Y}_1 = \mathbf{Y}_3, \mathbf{X}_2 = \mathbf{Y}_2$. By reassigning coordinates of vector $\mathbf{Y} : \mathbf{Y}_3 = \mathbf{X}_4, \mathbf{Y}_2 = \mathbf{X}_5, \mathbf{Y}_1 = \mathbf{X}_6$, we obtain the coordinate notation of the cluster structure in the chain of $N = 6$ elementary oscillators that appears based on synchronization of simple cells: $\mathbf{X}_1 = \mathbf{X}_3 = \mathbf{X}_4 = \mathbf{X}_6, \mathbf{X}_2 = \mathbf{X}_5$. Note that this notation also represents an integral manifold that occurs in the phase space of a lattice with 6 elements (see the beginning of this section).

Figure 6.10 shows results of the numerical experiment performed for system (6.10), the elements of which represent chaotic Chua's oscillators. The following parameter set has been used: $(\alpha, \beta, \gamma, m_0, m_1, \varepsilon) = (9.5; \ 14; \ 0.1; \ -1/7; \ 2/7; \ 0.215)$, and matrix $C = \mathrm{diag}(1, 0, 0)$.

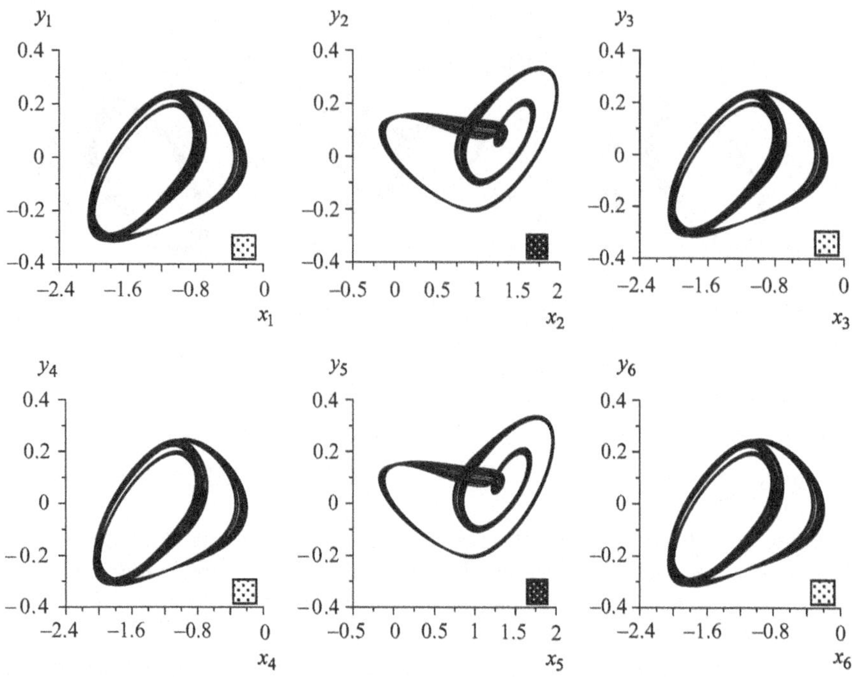

Fig. 6.10 Cluster structure in a chain of $N = 6$ elementary Chua's oscillators as a result of synchronization of two simple cells of O_a (2) type

As one can see, the cluster structure shown in Fig. 6.10 is "cut" into a pair of blocks—a pair of synchronized simple cells.

Let us summarize. First, the phenomenon of "cluster synchronization" represents the case of classical synchronization in the conventional understanding of this process and in this sense is not a new physical phenomenon. Second, in the case of systems with a small number of oscillators, two types of subjects have been identified, the classical synchronization of which generates cluster structures in a chain of dynamical systems. The considered, particular cases of cluster structures with a small number of elements can be naturally generalized to a general case. Namely: the first subject of cluster structures in chains is a cluster forming oscillator of the type O_s (n); the second one is a simple cell, a structured object that appears as a result of synchronization of a pair of cluster forming oscillators of the type O_a (n).

As it will be shown in the following sections, these two types of subjects exhaust all possible types of cluster structures in chains of oscillators, which represent the simplest form of lattices.

Note. The structure and role of O_s (n) and O_a (n) in formation of cluster structures involuntarily impose an analogy with quantum-mechanical quasiparticles: bosons and fermions. It is known that the former ones play an independent role in interactions, and the latter ones, before becoming subjects of interactions, were previously combined into pairs (they are born and "die" in pairs). Within the framework of these representations, it is possible to propose a quantum-mechanical interpretation of cluster dynamics of lattices.

Using properties of number N (divisibility) and properties of $O_s(n)$ and $O_a(n)$, it is possible to establish all of the integral manifolds in the phase space of a lattice and, similarly to the energy levels, to order them according to the number of clusters. As it will be shown below, most of such levels (or all levels) are metastable. This means that (in a typical case) when a portion of energy is transferred to oscillators of the lattice (initial conditions), which corresponds to the upper energy level, the interaction of particles (elementary oscillators) will lead to the origination of an ensemble of interacting quasiparticles, i.e. a cluster structure of a certain type that corresponds to the "upper" manifold. Since the energy level is unstable, then, after a while, this state of the lattice disappears. Due to the dissipative property of the system, a part of the energy is absorbed, i.e. the system goes from a lower energy level, to another metastable cluster structure and so on, up to the level that corresponds to integral manifold M_0. In general, such a scenario can resemble the process of transformation of a disturbed hydrodynamic flow into a laminar state through a sequence of successive structures.

6.2 Fusion and General Properties of Circuits of Cluster Structures

It is known that equivalent transformation in electrical circuits is a widely used method for analyzing and calculating electrical schemes, and this method is especially effective when these circuits are cumbersome and, therefore, are not widely available for these procedures in their original form. Let us remind the reader that the equivalent transformations of an electrical circuit are those transformations of some part of its electrical circuit, under which all currents of the chain segments, as well as the potentials of all points outside the transformed part, remain unchanged during the transformation process. This has a direct relation on the circuits of cluster structures for the following reasons (simple cases have already been discussed in the previous section). First, the lattices, the elements of which represent radio-technical self-generators (a superconductive junction is one of the examples), are also the subject of interest. Second, when analyzing cluster structures in the lattices of another physical nature, it is always possible to compare them with a certain radio-technical analogue. In this sense, any circuit of a cluster structure can be considered as a block diagram of a hypothetical grid of radio-technical self-generators, where the individual dynamic modes of these elements of the circuit are reflected by colors.

It must be said that the nature of the cluster structures that we have revealed (i.e. the synchronization of generalized oscillators (cluster forming oscillators) and simple cells) actually predetermines rules for the fusion of circuits of cluster structures, properties of these circuits, and their equivalent transformations, which are valid regardless of the geometric dimension of the lattices and their shape.

Let us provide a list of verbal definitions, general principles of the internal arrangement as well as elementary (equivalent) transformations of circuits of cluster structures.

1. Two directly connected, synchronized elementary oscillators, depicted in the diagram of the cluster structure using the same color, will be called the *cut pair*. The coupling between such oscillators can be removed by means of cutting (in the case of electric circuits, this corresponds to cutting the de-energized sections of the circuit representing equipotential points). On the contrary, any passive coupling can be established between same-colored schematic images and, in particular, a short-circuit can be made.

2. If a circuit has at least one cut pair, then we will call it the *cut circuit*, and otherwise the system is called the *non-cut circuit*.

3. The circuit of the cluster structure of a simple cell is non-cut.

4. If a certain circuit does not have any cut pairs, then it is either a circuit of a simple cell or a circuit of a separate cluster forming oscillator, or this circuit is not correct.

5. The procedure for combining two or more self-colored images by bending the circuit through the axes of symmetry or in another way will be called the *convolution of the circuit* (for the electric circuit, this corresponds to the procedure for connecting the equipotential points).

6. A circuit of a simple cell can be convoluted to the circuit of its basic cluster forming oscillator; that is, to a circuit that does not contain synchronized (same-colored) elementary oscillators.

7. Any circuit of a cluster structure, if it is not a circuit of a single cell or a circuit of a separate cluster forming oscillator, during the cutting of all cut pairs, is decomposed into identical blocks depicting the basic cluster forming oscillator (boson) or a circuit of the basic simple cell (fermion).

On the basis of these simple principles and equally simple elementary transformations of circuits (corresponding to the level of school physics), we will establish all possible types of cluster structures in a chain and in a ring of elementary oscillators. Also, mathematically non-trivial statements concerning the completeness of the aggregates of these types of structures will be proved.

Let us turn to the fusion of circuits of cluster structures. Like the connection of electric conductors, the connection of circuits of cluster forming oscillators as parts of the complete circuit of the cluster structure, depicted in Fig. 6.11, we will call the sequential coupling (sequential fusion of a circuit of a cluster structure).

Matrix $\mathbf{D} = \mathrm{diag}(d_{11}, d_{22}, \ldots d_{ss})$, in which the diagonal elements d_{ii} equal either 0 or 1, will be called the fusion matrix of the cluster structure. The subscripts of the unit elements determine the number of coupled elementary oscillators as elements of the cluster forming oscillator or simple cell. The dimension of this matrix is equal to the number of oscillators forming the cluster forming oscillator ($m = n$, see Fig. 6.11a) or simple cell ($m = 2n - 1$, see Fig. 6.11b). The matrix of sequential fusion is matrix $\mathbf{D} = \mathbf{C}^* = \mathrm{diag}(1, 0, \ldots 0)$ or $\mathbf{C}_* = \mathrm{diag}(0, 0, \ldots 1)$, depending on

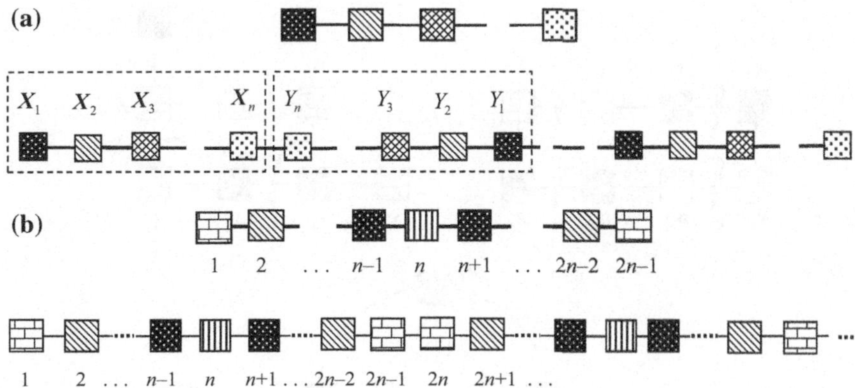

Fig. 6.11 A circuit (scheme) of cluster forming oscillator $O_s(n)$ and sequential fusion of cluster structure on its basis (**a**); an image of a simple cell of cluster forming oscillator $O_a(n)$ and sequential fusion of a cluster structure on its basis (**b**)

the order of the zero-vectors of elementary oscillators that represent the coordinates of the vectors of the cluster forming oscillator or simple cell.

The coupling shown in Fig. 6.12 will be called the parallel coupling.

In this particular case, the fusion matrix $\mathbf{D} = \mathrm{diag}(1, 1, \ldots, 1) = \mathbf{I}$ is a unit matrix.

All other types of fusion of block schemes from identical circuit-blocks (cluster forming oscillator or simple cell) are classified as a mixed type (see Fig, 6.13).

Once again, we pay attention to the features of existence and coexistence of cluster structures in homogeneous lattices: if a synthesized circuit of the cluster structure uniquely determines the lattice itself, then, in a general case, this does not mean that the given structure will be unique in the resulting material lattice. A synthesized structure, as well as its accompanying structures, are associated with certain initial conditions of the obtained lattice. Thus, if the initial conditions of the lattice are given in accordance with the circuit (identical or sufficiently close to the identical initial conditions for same-colored oscillators), then its dynamics will continue to follow

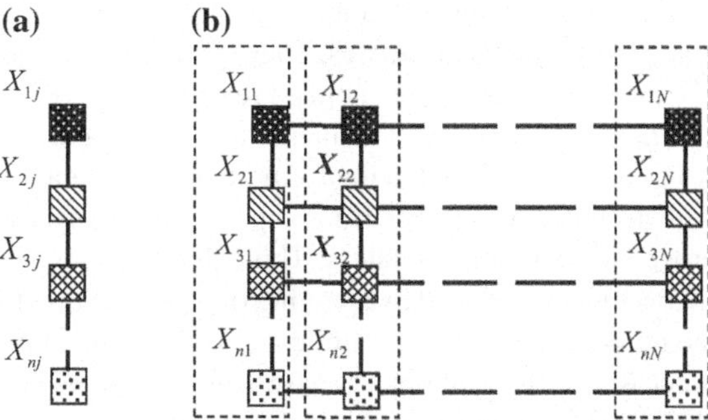

Fig. 6.12 A circuit (scheme) of cluster forming oscillator $O_s(n)$ (**a**); a parallel fusion–tape-like cluster structure in a two-dimensional lattice (**b**)

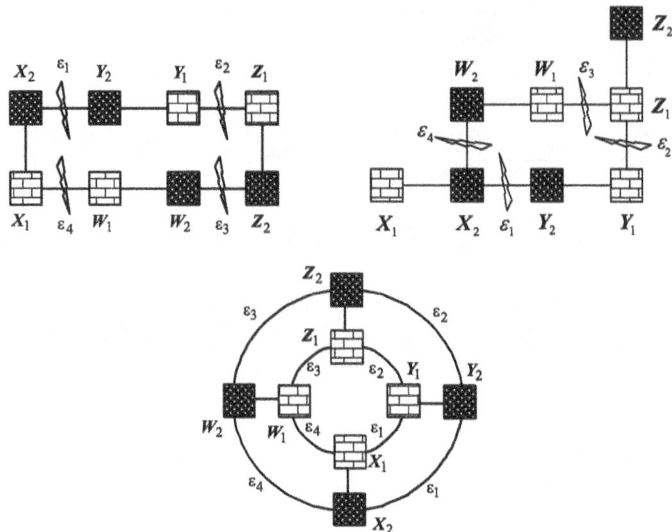

Fig. 6.13 Examples of a mixed type of fusion of cluster structure based on cluster forming oscillator $O_s(2)$

in accordance with the synthesized circuit. Under other initial conditions, the result may be different. We can say that structures in homogeneous lattices are determined by the ordered inhomogeneity of the initial conditions.

As a result of the aforesaid matter, questions arise about the fusion of structures in lattices by introducing a predetermined ordered inhomogeneity into the material part of the lattice. All problems of the fusion of orderly inhomogeneous lattices with the corresponding cluster dynamics have a positive solution, and it lies within the framework of the same principles and techniques as in the case of homogeneous lattices. It is sufficient to assume that the operating mode of the cluster forming oscillator (attractor) is caused not by the initial conditions, but by the inhomogeneity of the parameters of the elementary oscillators that form it. A synthesized cluster structure in such a lattice may have global stability and controllability, including a remote control.

Here are two more examples of "clustering" of rather non-ordinary objects—objects with fractal properties. They represent one of the most outstanding directions of modern theoretical and applied research [17, 18].

The Cayley Tree. As a mathematical object, the Cayley tree is formed as follows. A figure is constructed in the form of an equilateral letter "H". Then a similar figure with a similarity factor of 1/2 is superimposed on each of the ends of this figure. The process of building is endless. Three iterations are shown in Fig. 6.14a. In practice, such a tree is produced by spraying or etching a thin film [19]. Naturally, in practice, the number of iterations is finite and is determined by the level of the technological process, i.e. physical objects have a pseudofractal structure, but their properties are close to the properties of the real fractals.

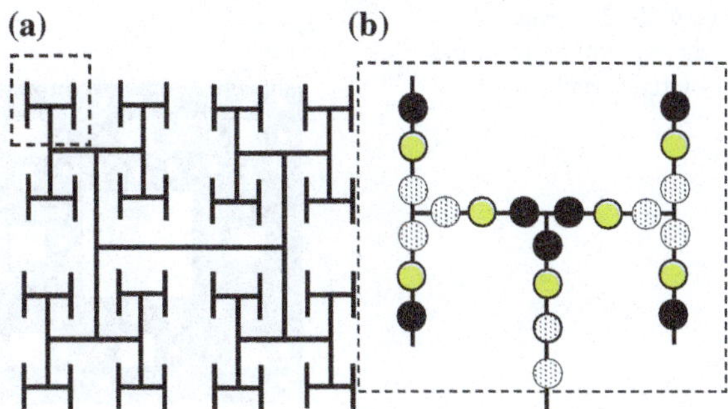

Fig. 6.14 The Cayley tree (**a**); and a part of the clusterized tree (**b**)

Let us assume that technology allows us to construct some active nonlinear elements (conditional oscillators) of sufficiently small sizes. In this case, taking as a basis a circuit of some cluster forming oscillator of a chain with the number of elements n or a circuit of a simple cell with $2n - 1$ elements, we "lay out" this circuit (according to the rules of the fusion) on each branch of the smallest element of the tree (in Fig. 6.14, it is indicated by a dashed line) the required number of times. Thus, we determine the necessary number of conditional oscillators to be placed on each branch of the tree. In the same way, the synthesis of the structure extends to the entire tree. As a result, we obtain a "clustered" tree, a segment of which is shown in Fig. 6.14b, which depicts the structure based on $O_s(3)$. However, other cluster forming oscillators or simple cells could be chosen as well.

Now let us assume that all of the conditional oscillators are identical. In this case, we will deal with a homogeneous lattice, although differently than for previous examples by its over-the-original form. The physics remains the same. We know that, along with the constructed cluster structure, other structures are potentially possible in this lattice. For this reason, in order to realize a necessary process, it is necessary to have the technical ability (a skill) to define specific initial conditions of the lattice (before it is put into operation). In addition, in the case of instability of this structure, additional measures are required to maintain its metastable state. Since control of the initial conditions represents an almost irresolvable task, it is most likely the case that the entire task is useless. A completely different situation can arise if the oscillators are controlled by some parameter like, for example, the frequency or phase of the oscillations. In this case, we will deal with an orderly inhomogeneous lattice, in which the cluster dynamic regime, on the one hand, will be stable, and on the other hand will have controllable parameters. In addition, controlling the parameters of the oscillators will create a possibility of switching the lattice from one cluster structure to another one. It could be possible to, in the case of antennas with a fractal lattice, obtain directional patterns with new unusual properties using the same approach.

The Sierpinsky carpet. This object is constructed as follows: a square is divided into nine equal squares; the middle square is removed, and with the remaining squares the division procedure is repeated. The process is endless. The resulting shape represents a so-called Sierpinsky carpet. Three iterations of a square are shown in Fig. 6.15a. The unfinished carpet is marked by a black color, and a white color shows the removed parts of the square.

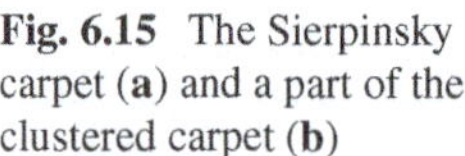

Fig. 6.15 The Sierpinsky carpet (**a**) and a part of the clustered carpet (**b**)

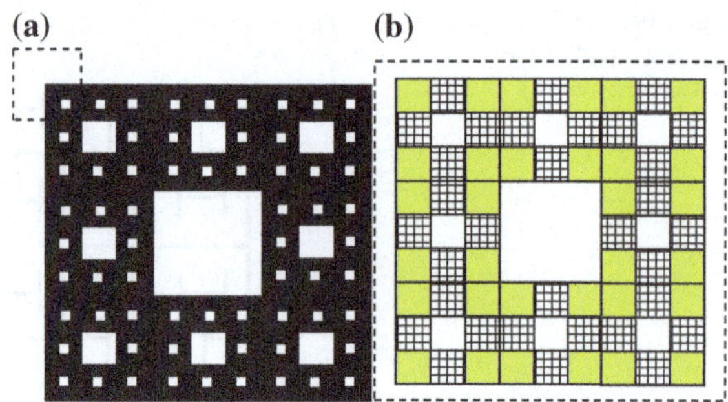

As in the previous example, we choose the smallest element of the carpet—the square, which is highlighted in Fig. 6.15 with a dotted line, as well as a diagram of the required cluster forming oscillator or simple cell. The dimensions of the conditional oscillators as the "body" of the lattice, as determined by the possibilities of the manufacturing technology, must correspond to each other. Figure 6.15b shows an example of the base element: a circuit of the square simple cell (3 × 3) with two clusters. Following the rules, we synthesize a circuit of the cluster structure, first in this element and then on the entire surface of the carpet. A fraction of the cluster structure is shown in Fig. 6.15b. The couplings of the elementary oscillators are represented by the contact of the sides of the squares. The properties of the obtained structures in homogeneous and non-homogeneous lattices are discussed above.

During the fusion of inhomogeneous lattices with a given cluster structure, the elementary oscillators that comprise the cluster forming oscillator and, respectively, the simple cell, can differ not only in parameters but also in types. This does not change the above rules for the fusion of circuits of cluster structures.

Let us provide an example illustrating what has been said above. Figure 6.16 shows individual phase portraits of three oscillators representing "candidates" for the cluster forming oscillator O_s (3). The first and the second ones are Chua's oscillators with parameters $(\beta, \gamma, m_0, m_1) = (14; \ 0.1; \ -1/7; \ 2/7)$ and $\alpha_1 = 9$, $\alpha_2 = 9.5$, respectively. The third one is the Lorenz oscillator with parameters $\sigma = 10$, $r = 24$ and $b = 1$.

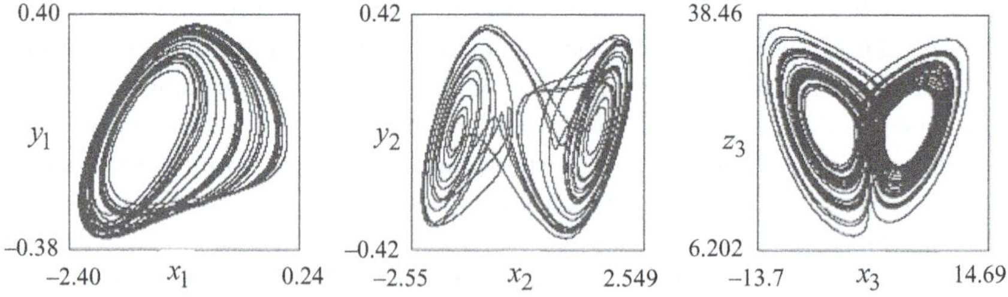

Fig. 6.16 Chaotic attractors of the oscillators before origination of the cluster forming oscillator O_s (3)

Figure 6.17 shows projections of a cluster attractor of cluster forming oscillator O_s (3) onto coordinate planes. The matrix of couplings of elementary oscillators has the form $\varepsilon C = 0, 5I$. Figure 6.18 shows a regime of synchronization of a pair of cluster forming oscillators with matrix and scalar parameter of coupling $\varepsilon C_* = 25\mathrm{diag}(0, 0, 1)$.

A visual comparison of the phase portraits of the oscillators is sufficient to state that we are dealing with the synchronization of a pair of cluster forming oscillators. The synchronization is stable.

In conclusion of this section, we would like to show an example from biology [20].

Being unfolded, a DNA molecule has the form of stairs consisting of separate links (in fact, it is twisted in a helix). The number of links is $\sim 10^5$. A segment of such a ladder is shown in Fig. 6.19a. In this figure, the following notations have been

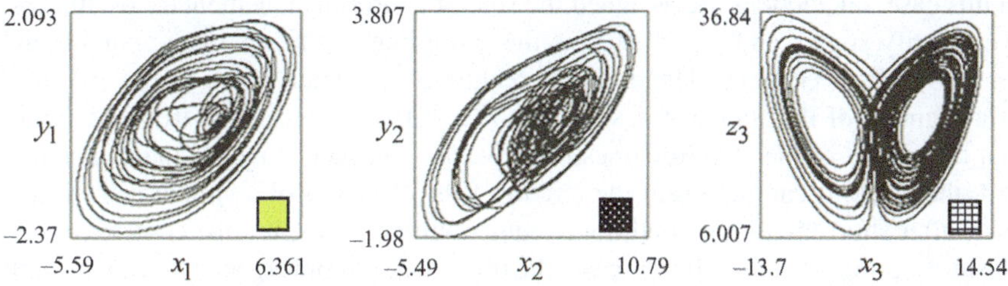

Fig. 6.17 Projections of cluster attractor onto the coordinate planes

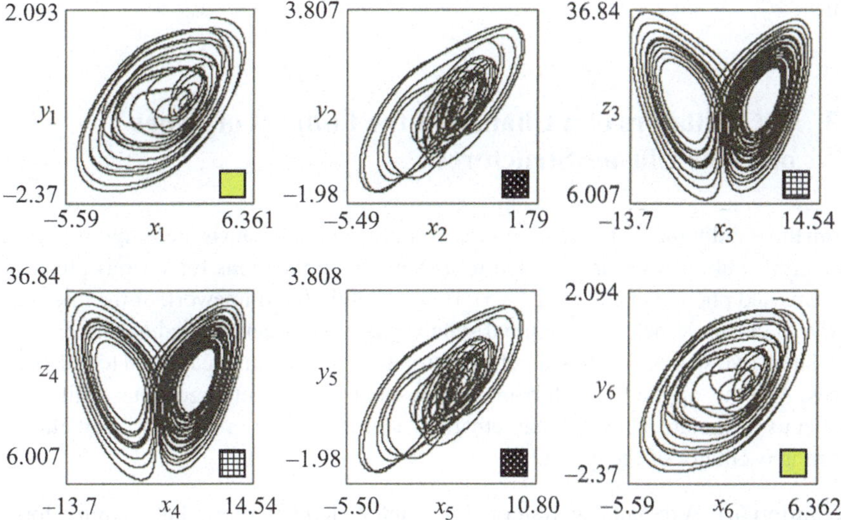

Fig. 6.18 Cluster structure as a synchronization of pair of cluster forming oscillators of the type O_s (3)

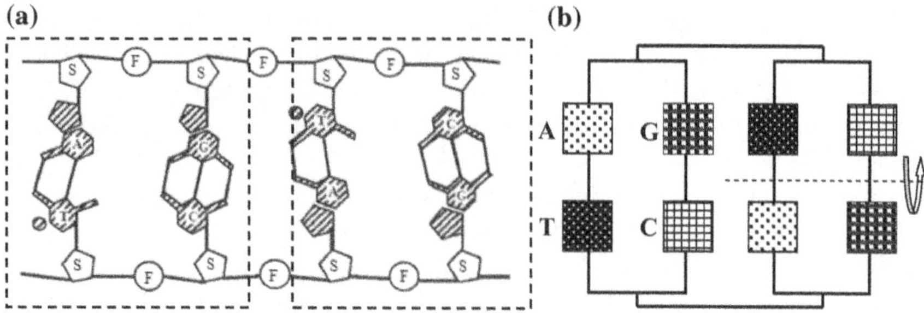

Fig. 6.19 A segment of the unfolded DNA molecule

used: S denotes sugar molecules; F denotes molecules of phosphoric acid; T, A, G, and C denote nucleotides: thymine, adenine, guanine, and cytosine, respectively [21]. In this case, nucleotides are assigned the role of conditional elementary oscillators. We classify sugars and phosphates as the "coupling" elements of the conditional oscillators T, A, G, and C. Under these conditions, a scheme of the cluster structure of a segment of the staircase is shown in Fig. 6.19b. One can see from the figure that the circuit of the structure does not have any cut pair. It means that it is uncut and, therefore, it can represent the circuit of the structure of a simple cell. In this case, after short-circuiting conditional equipotential points, it must contract to one cluster forming oscillator. In this case, the procedure of convergence of a simple cell geometrically corresponds to the overlapping of two blocks after the rotation of the second one by 180 degrees. Thus, the block (T, A, G, C) represents a cluster forming oscillator, and the structure shown in Fig. 6.19a is the cluster structure of a simple cell.

6.3   C-Oscillators of a Chain and the Fullness of Types of Their Cluster Structures

From the standpoint of cluster structures, a homogeneous chain is the simplest object. Numerous publications are devoted to its various applications for various physical systems and phenomena [3, 5, 22–25]. Below, within the framework of the interpretation of cluster structures as mutual synchronization of generalized oscillators, we will show (we will prove) that all cluster structures in a chain are limited to only two types. And this applies to both homogeneous and orderly heterogeneous chains.

Let us formulate a general mathematical definition of the aforementioned cluster forming oscillator of the type $O_s(n)$.

Definition 6.1 A dynamical system of n coupled elementary oscillators of the form

$$\dot{\mathbf{V}} = \mathbf{H}(\mathbf{V}), \tag{6.11}$$

$$\mathbf{V} = (\mathbf{X}_1, \mathbf{X}_2, \ldots, \mathbf{X}_n)^T,$$

$$\mathbf{H(V)} = (\mathbf{F}_1(\mathbf{X}_1), \mathbf{F}_2(\mathbf{X}_2), \ldots, \mathbf{F}_n(\mathbf{X}_n))^T - \varepsilon(\mathbf{B}_s \otimes \mathbf{C})\mathbf{V},$$

$$\mathbf{B}_s = \begin{pmatrix} 1 & -1 & 0 & \cdots & \cdots & 0 & 0 \\ -1 & 2 & -1 & \ddots & \cdots & 0 & 0 \\ 0 & -1 & 2 & \ddots & \ddots & 0 & 0 \\ \vdots & \ddots & \ddots & \ddots & \ddots & \ddots & \vdots \\ \vdots & \vdots & \ddots & \ddots & 2 & -1 & 0 \\ 0 & 0 & 0 & \ddots & -1 & 2 & -1 \\ 0 & 0 & 0 & \cdots & 0 & -1 & 1 \end{pmatrix}$$

Is called a C-oscillator of the type $O_s(n)$ conditioned that in the phase space $G(\mathbf{V})$ there exists attractor $A_s(n)$, such that for $\forall(\mathbf{X}_1, \mathbf{X}_2, \ldots, \mathbf{X}_n) \in A_s(n)$, $\mathbf{X}_i \neq \mathbf{X}_j$, $i \neq j, i, j = 1, 2, \ldots, n$. $\mathbf{B} \otimes \mathbf{C}$ is the coupling matrix of elementary oscillators.

Elementary oscillators in (6.11) can be either identical, which corresponds to a homogeneous chain, or not identical by their parameters or even by their nature. The coupling of the oscillators in (6.11) is assumed to be homogeneous and local, although this circumstance is not fundamental and does not change the general essence of the matter.

By writing system (6.11) coordinate-wise, it is easy to verify that it corresponds to a segment of a chain with Neumann boundary conditions, which are satisfied "automatically" since they are embedded in the coupling matrix. For convenience, Fig. 6.20 shows once again a scheme of the cluster forming oscillator of the type $O_s(n)$ in its "operational" dynamical regime [on attractor $A_s(n)$], while Fig. 6.21 shows a process of fusion of a circuit of the cluster structure in a chain based on this cluster forming oscillator.

By analogy, let us formulate the definition of a cluster forming oscillator of the type $O_a(n)$.

Definition 6.2 We call a system of coupled elementary oscillators of the form

Fig. 6.20 A circuit of $O_s(n)$

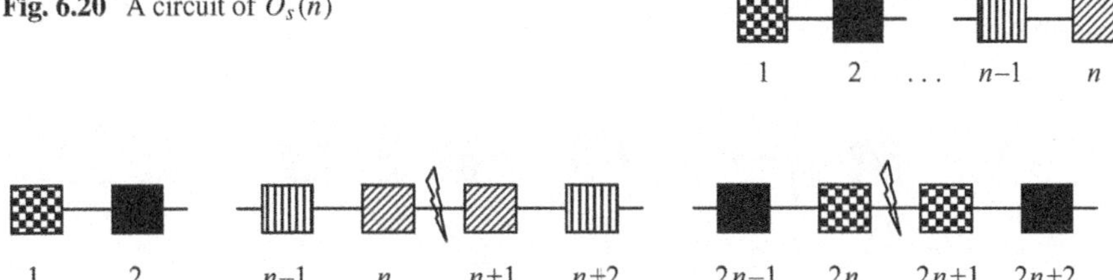

Fig. 6.21 The fusion of a circuit of cluster structure based on $O_s(n)$

$$\dot{\mathbf{V}} = \mathbf{H}(\mathbf{V}), \tag{6.12}$$

$$\mathbf{V} = (\mathbf{X}_1, \mathbf{X}_2, \ldots, \mathbf{X}_n)^T,$$

$$\mathbf{H}(\mathbf{V}) = (\mathbf{F}_1(\mathbf{X}_1), \mathbf{F}_2(\mathbf{X}_2), \ldots, \mathbf{F}_n(\mathbf{X}_n))^T - \varepsilon(\mathbf{B}_a \otimes \mathbf{C})\mathbf{V},$$

$$\mathbf{B}_a = \begin{pmatrix} 1 & -1 & 0 & \cdots & \cdots & 0 & 0 \\ -1 & 2 & -1 & \ddots & \cdots & 0 & 0 \\ 0 & -1 & 2 & \ddots & \ddots & 0 & 0 \\ \vdots & \ddots & \ddots & \ddots & \ddots & \ddots & \vdots \\ \vdots & \vdots & \ddots & \ddots & 2 & -1 & 0 \\ 0 & 0 & 0 & \ddots & -1 & 2 & -1 \\ 0 & 0 & 0 & \cdots & 0 & -2 & 1 \end{pmatrix}$$

The C-oscillator of the type $O_a(n)$ is conditioned that in the phase space $G(\mathbf{V})$ there exists attractor $A_a(n)$, such that for $\forall(\mathbf{X}_1, \mathbf{X}_2, \ldots, \mathbf{X}_n) \in A_a(n)$, $\mathbf{X}_i \neq \mathbf{X}_j$, $i \neq j, i, j = 1, 2, \ldots, n$. $\mathbf{B}_a \otimes \mathbf{C}$ is the coupling matrix of elementary oscillators.

The sense of the variables and parameters of Eq. (6.12) is the same as explained above.

Recall that a cluster forming oscillator of this type exists only in pairs. The mutual synchronization of such a pair in the limiting (rough) case of the connection of two cluster forming oscillators to each other determines the cluster structure of the simple cell. In turn, the formed simple cells are the potential subjects of cluster structures in the chain. We note that after a rigid coupling of the pair of terminal elementary oscillators of the cluster forming oscillator, they actually merge and, as a consequence, degenerate in the number of degrees of freedom of the coupled system by the number of degrees of freedom of one elementary oscillator. In this case, the number of acting elementary oscillators in a simple cell becomes equal to $2n - 1$. Equations of a simple cell are the equations of a segment of the chain with Neumann boundary conditions and number of elementary oscillators of $2n - 1$. A fusion of a circuit of a cluster structure of a simple cell can be interpreted as an addition to circuit $O_a(n)$ of its mirrored image (Fig. 6.22). The bold vertical line in the figure symbolizes a mirror. Figure 6.23 shows a fusion of a circuit of a cluster structure in a chain based on a circuit of the simple cell of the type $O_a(n)$.

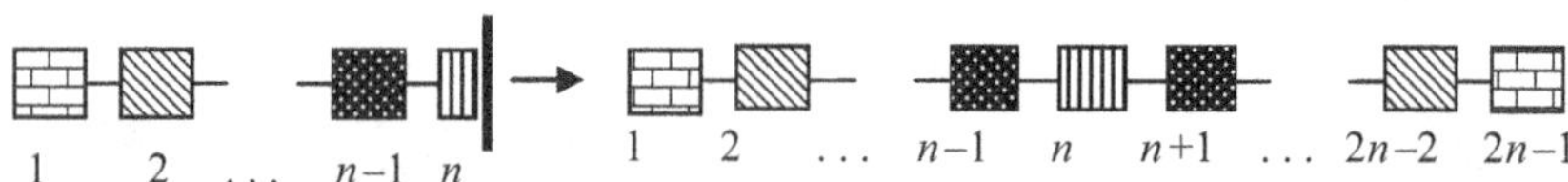

Fig. 6.22 A circuit of the cluster forming oscillator of type $O_a(n)$ and a fusion of a circuit of a cluster structure of a simple cell on its base

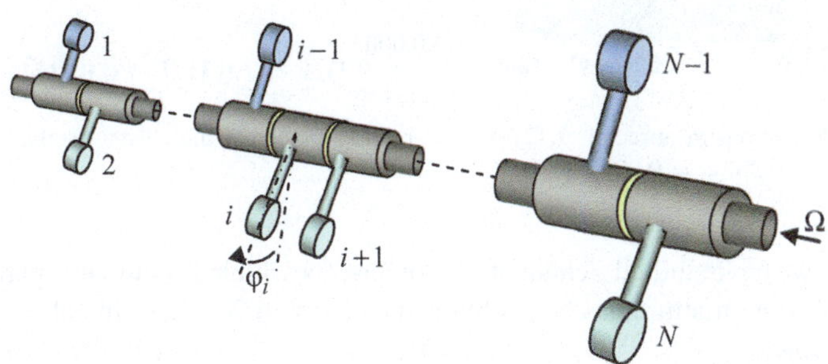

Fig. 6.23 A fusion of the circuit of a cluster structure based on a circuit of a simple cell of the type $O_a(n)$

Fig. 6.24 A chain of Froude pendula with structured dynamics

As already mentioned regarding to the general principles of the theory of cluster structures, the division of systems into oscillators and rotators does not make a difference. The significance of this division appears only in the study of specific systems and their attractors, when systems of rotators exhibit specific features associated with the cylindricality of the phase space. For this reason, in what follows, we keep the adopted terminology and notation. Speaking of oscillators, we will bear in mind that the same applies to rotators as well. Let us just add to the aforesaid matter an illustration of a numerical experiment for cluster structures in a chain of rotators, a physical example of which is shown in Fig. 6.24.

We have numerically studied a chain of rotators with the following vectors of dynamical state of elements: $\mathbf{X}_i = (\varphi_i, x_i)^T$ and $\mathbf{F}(\mathbf{X}_i) = (x_i, -\lambda x_i - \sin \varphi_i + \gamma + A \sin \omega_0 \tau)^T$ with scalar coupling parameter σ and matrix $\mathbf{C} = \begin{pmatrix} 0 & 0 \\ 0 & 1 \end{pmatrix}$. In particular, such a system models dynamics of a chain of Froude pendula shown in Fig. 6.24.

1. *Cluster structures of autonomous chain.* It is not difficult to demonstrate examples of cluster structures in a chain of rotators. For instance, in parameter domain (2) (see Fig. 1.1), a single rotator, depending on the initial conditions, may either be in equilibrium or in rotational motion. This means that, in the phase space of a pair of coupled rotators, there exists a stable oscillatory-rotational limit cycle (at least, for a weak coupling, this is guaranteed by the continuity of solutions by parameter). This limit cycle represents one of the cluster attractors of the system (see Fig. 6.25). The first of the coupled pendula rotates, while the second one experiences slight oscillations in the vicinity of the lower equilibrium.

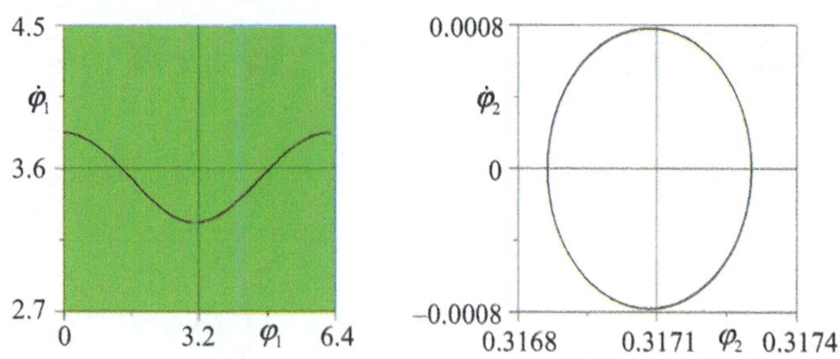

Fig. 6.25 Projections of attractor $A_s(2)$ onto coordinate planes of the cluster oscillator $O_s(2)$: $\lambda = 0.07$, $\gamma = 0.28$, $\sigma = 0.09$

Further, we have studied a chain of 8 elements, for which, for the same parameters, the same cluster structure has been obtained (see Fig. 6.26). Looking ahead, we say that it is stable.

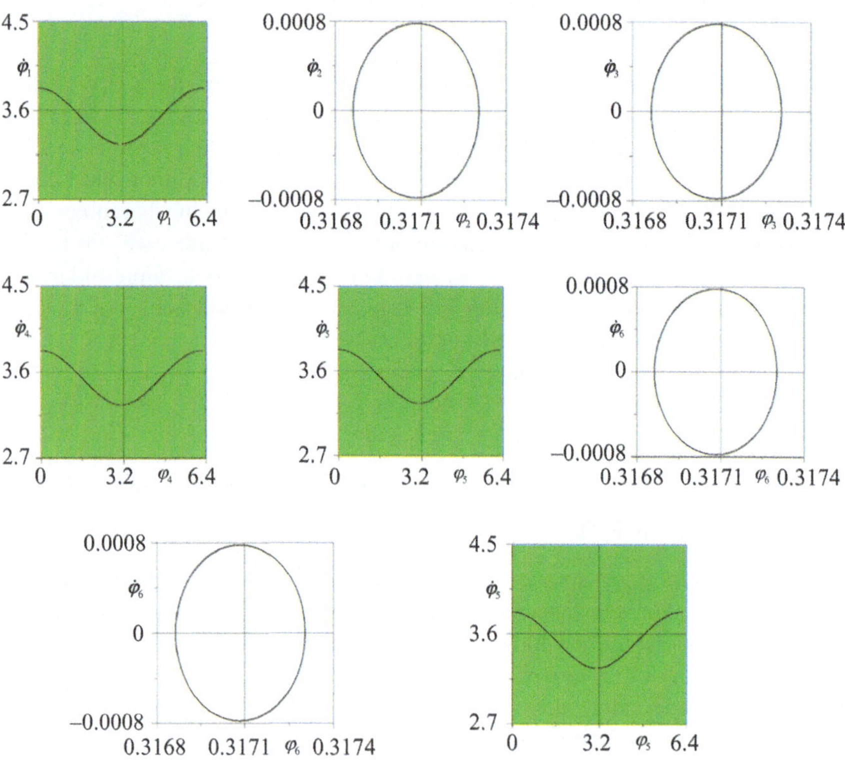

Fig. 6.26 Cluster structure based on synchronization of four cluster forming oscillators of $O_s(2)$ type in a chain of $N = 8$ elementary rotators

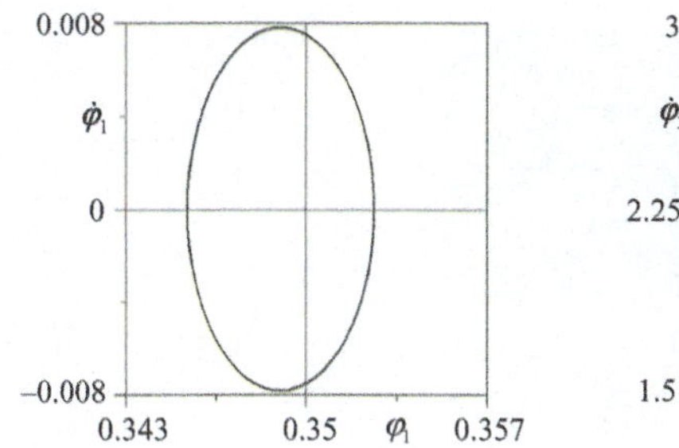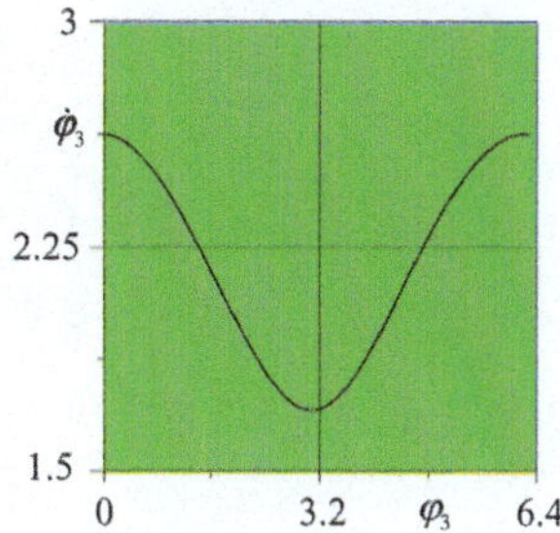

Fig. 6.27 Projections of attractor $A_a(2)$ into same-titled coordinate planes of the cluster forming oscillator of the type $O_a(2)$

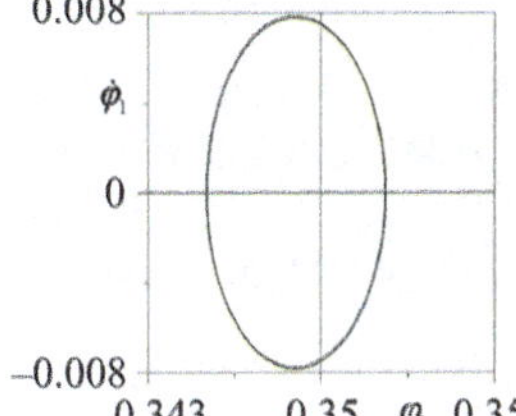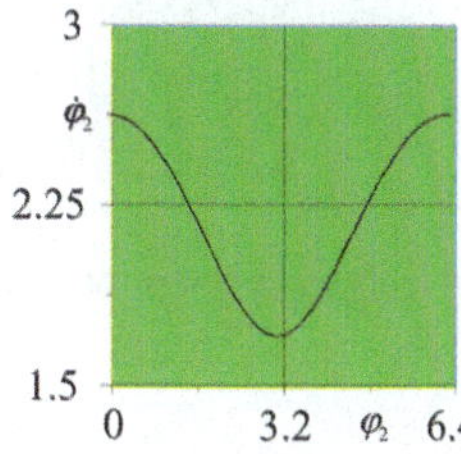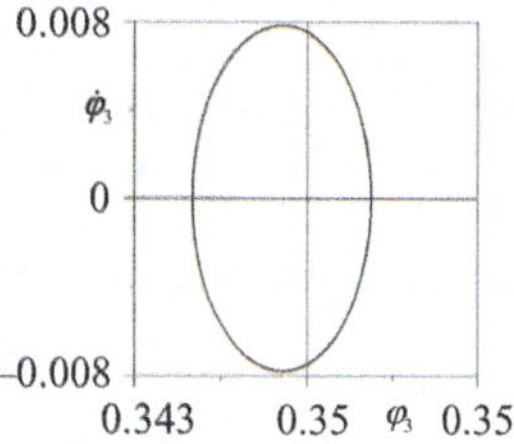

Fig. 6.28 Cluster structure of a simple cell of cluster forming oscillator $O_a(2)$

For parameters of the individual rotators from parameter domain (2) (see Fig. 1.1), a cluster forming oscillator of the type $O_a(2)$. has been modeled according to Definition 6.2. Its attractor $A_a(2)$, with the parameters of the coupled system $\lambda = 0.07$, $\gamma = 0.28$, $\sigma = 0.29$, is shown in Fig. 6.27. The similarity with attractor $A_s(2)$ is only in appearance, but they are fundamentally different.

Figure 6.28 shows a cluster structure of a simple cell of cluster forming oscillator $O_a(2)$ (stable), which has been obtained for the same parameters in a chain of three elements.

Further, for the same parameters, we have studied a chain of 6 elements, in which a cluster structure shown in Fig. 6.29 has been obtained.

2. *Cluster structures of non-autonomous chain.* In this case, to select parameters that are necessary for the existence of cluster attractors, we turn to the bifurcation diagram shown in Fig. 1.7a. As one can see from the figure, parameters from all domains suffice except for parameter domain (1): for these parameters a globally stable in-phase synchronization occurs, while we are looking for different partial dynamics of rotators. An example of cluster attractor $A_s(2)$ is shown in Fig. 6.30. The following parameter set has been used: $\lambda = 0.07$, $\gamma = 0.51$, $\sigma = 0.079$, $A = 0.28$, $\omega = 0.56$.

In this case, a cluster attractor is a three-dimensional torus T^3, which corresponds to the regime of beatings of rotators with each other as well as with the periodic

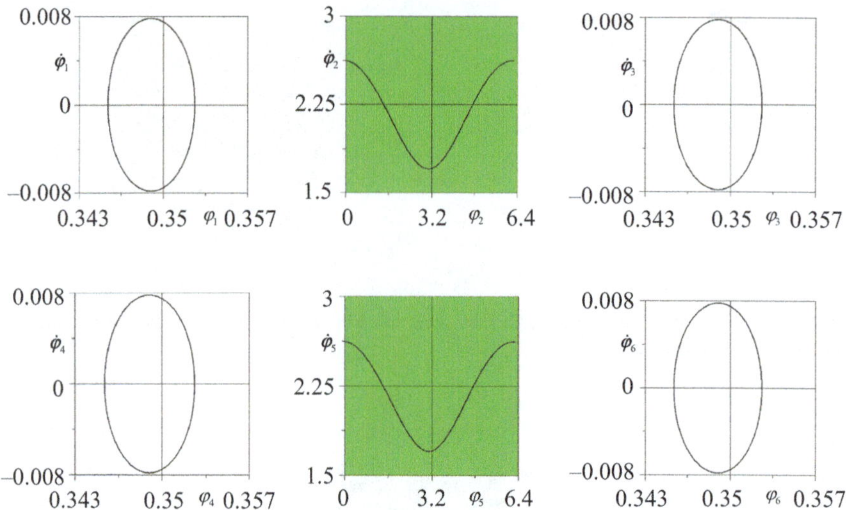

Fig. 6.29 Stable cluster structure as a synchronization of two simple cells of the type $O_a(2)$

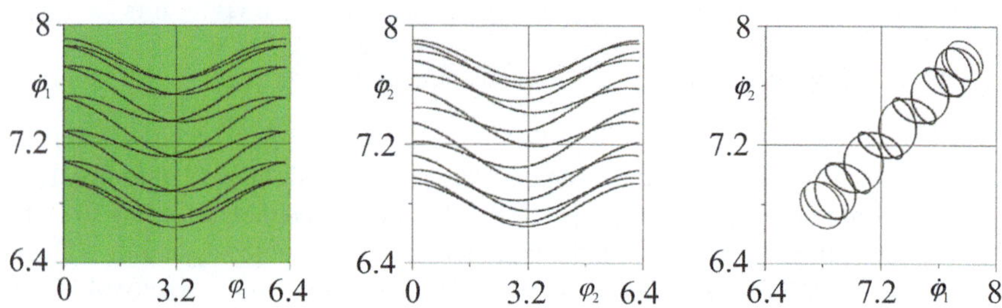

Fig. 6.30 Projections of attractor $A_s(2)$ into same-titled and cross-coordinate planes of cluster forming oscillator $O_s(2)$

external force, which means the absence of both mutual and master-slave synchronization. In what follows, without going into much detail, we show results of the numerical experiment (Figs. 6.31, 6.32, 6.33, 6.34). All cluster structures are stable.

Now let us subdivide all cluster structures in a lattice into different types. A type of set of structures will be determined according to the type of basic cluster forming oscillator that defines each structure of this set, i.e. the number of types of structures in a lattice is the number of all types of cluster forming oscillators that are possible in this lattice.

Theorem 6.1 *$O_s(n)$ and $O_a(n)$ constitute the full set of types of cluster forming oscillators in a chain of elementary oscillators with Neumann boundary conditions. Synchronization of these cluster forming oscillators and their simple cells defines all possible types of cluster structures existing in the chain* [26].

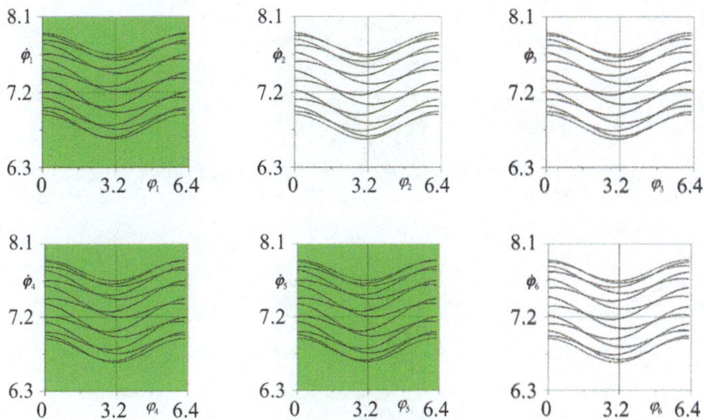

Fig. 6.31 Stable cluster structure based on the synchronization of three cluster forming oscillators of the type $O_s(2)$

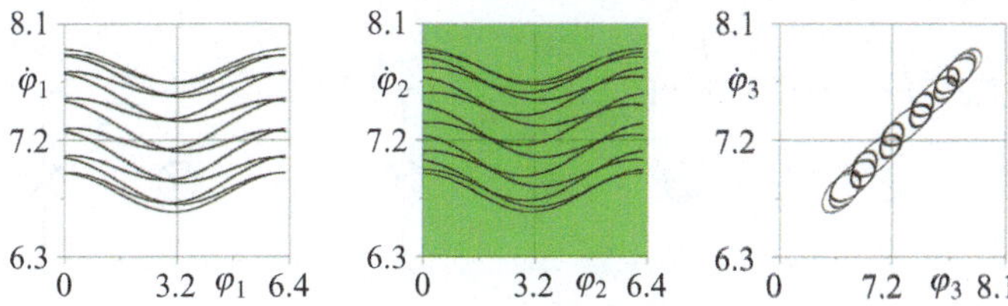

Fig. 6.32 Projections of attractor $A_a(2)$ onto the coordinate planes of cluster forming oscillator $O_a(2)$. The following parameter set has been used: $\lambda = 0.07$, $\gamma = 0.51$, $\sigma = 0.079$, A = 0.28, $\omega = 0.56$

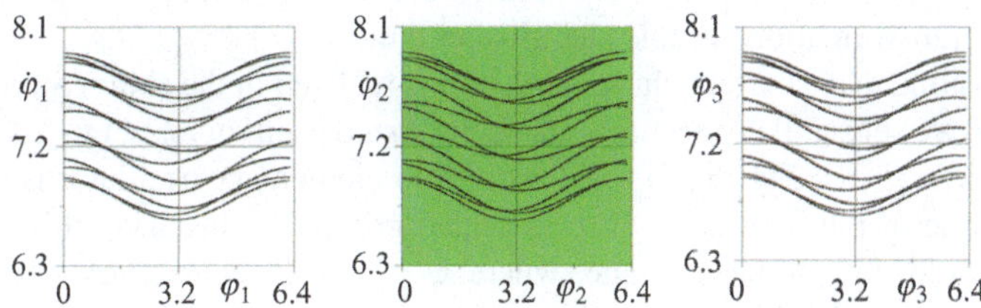

Fig. 6.33 Stable cluster structure of a simple cell $O_a(2)$

Proof We will use an electric scheme of the chain of oscillators (rotators) shown in Fig. 6.35. Let us rewrite equations for a chain in a common form

$$\dot{\mathbf{X}}_i = \mathbf{F}_i(\mathbf{X}_i\,,\,t) + \varepsilon\mathbf{C}(\mathbf{X}_{i-1} - 2\mathbf{X}_i + \mathbf{X}_{i+1}) \tag{6.13}$$

With the boundary condition at the left end $\mathbf{X}_0 = \mathbf{X}_1$. The boundary condition at the right end will be formulated during the proof.

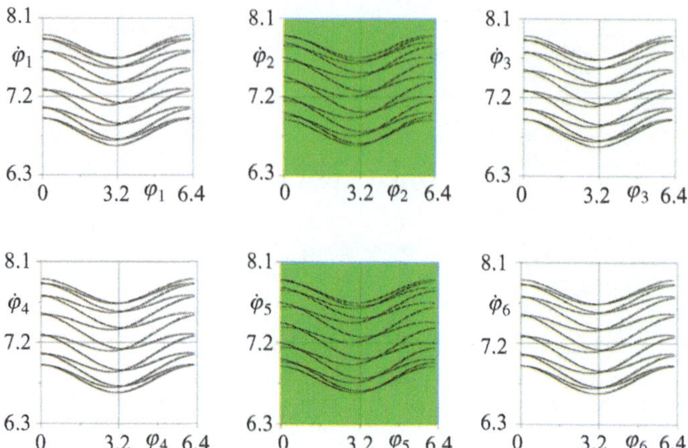

Fig. 6.34 Stable cluster structure based on the synchronization of two simple cells $O_a(2)$

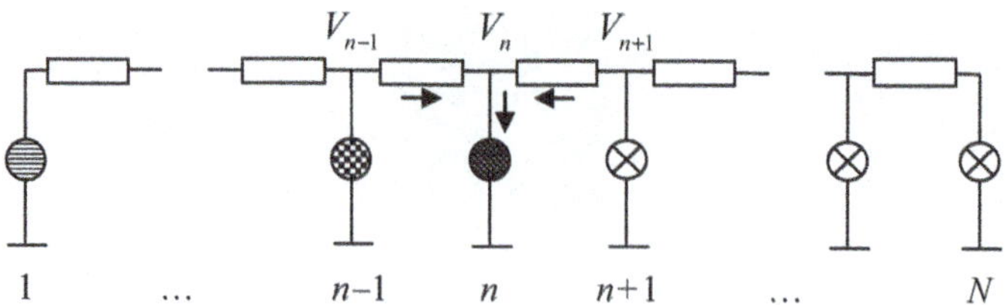

Fig. 6.35 A chain of oscillators (rotators)

For the proof, we use equivalent transformations of circuits of structures as elementary transformations of this electrical circuit.

Let us assume that a certain cluster structure is realized in the chain. Starting with the first one, we select all consecutive unsynchronized oscillators. Let us assume that their number is n (see Fig. 6.35), i.e. the $(n + 1)$ elementary oscillator is synchronized with one of the oscillators of the selected group. For instance, this oscillator is synchronized with some k-th one, where $k \leq n - 2$, elementary oscillator of the selected group. In this case, values of all same-titled variables of this pair of oscillators are identical at any moment of time. The points at their "inputs" are equipotential: $V_k = V_{n+1}$. We perform equivalent circuit transformations that correspond to the transformation of the circuit of the cluster structure: after connecting (short-circuiting) equipotential points with a bridge, we perform a sequence of transformations, as shown in Fig. 6.36.

Let us write equations for the electric currents of synchronized oscillators in the first and last positions of the circuit, respectively:

$$V_n - 2V_{n+1} + V_{n+2} = IR, \quad V_{k+1} - 2V_k + V_{n+2} = IR.$$

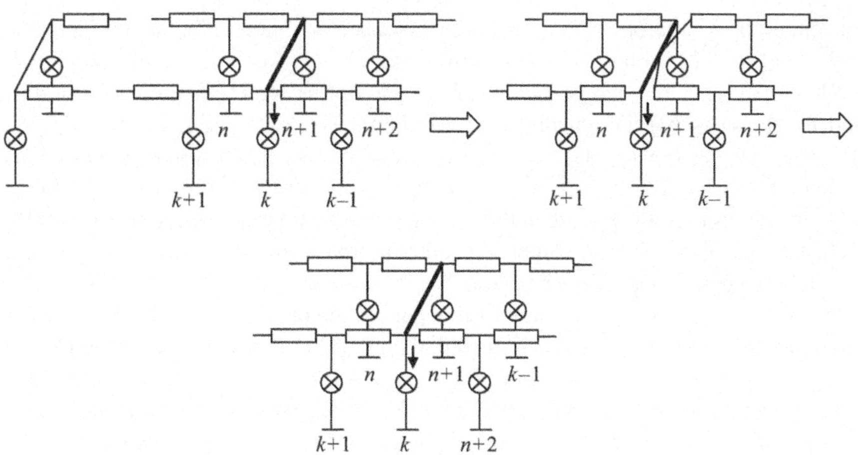

Fig. 6.36 Equivalent transformations of the chain

From these equations, we obtain $V_n = V_{k+1}$. Except for cases $k = n$ and $k = n-1$, this contradicts the condition of the non-synchronization of the first n oscillators. Thus, the $(n + 1)$ oscillator cannot be synchronized with any $k \leq n - 2$ oscillator. Two remaining cases are: (1) $k = n$, (2) $k = n - 1$.

(1)   Suppose that $k = n$, so that we obtain $V_n = V_{n+1}$ and $I_n = I_{n+1}$. On the other hand, the following equations are valid for these oscillators: $V_{n-1} - V_n = I_n R$, $V_{n+2} - V_{n+1} = I_{n+1} R$. From here, we obtain $V_{n-1} = V_{n+2}$, i.e. the condition of the synchronization of the $(n - 1)$ and $(n + 2)$ oscillators. Writing the Kirchhoff equations for this pair of oscillators, we obtain: $V_{n-2} + V_n - 2V_{n-1} = I_{n-1} R$, $V_{n+3} + V_{n+1} - 2V_{n+2} = I_{n+2} R$, from which it follows that $V_{n-2} = V_{n+3}$ and the corresponding pair of oscillators is also synchronized.

Continuing the successive writing of equations for synchronized oscillators, we obtain $V_1 = V_{2n}$. Further, we obtain a "fork": the oscillator with number $(2n + 1)$ can be synchronized with the $2n$ one or with the $(2n - 1)$ one. Let us assume that it is synchronized with the $(2n - 1)$ one, so that $V_{2n+1} = V_{2n-1}$. From the equations for the first and $2n$ oscillators, we obtain: $V_2 - V_1 = I_1 R$, $V_{2n-1} + V_{2n} - 2V_{2n+1} = I_{2n} R$. Thus $V_2 = V_1$, but this contradicts the condition of absence of the synchronization for the first n oscillators.

Thus, the $(2n + 1)$ oscillator is synchronized with the $2n$ one. Repeating such kind of reasoning again and again, we obtain the order of synchronized oscillators in the chain, as shown in Fig. 6.21, i.e. the structure based on $O_s(n)$. We show that, in this case, the cluster structure contains an integer number of images of cluster forming oscillators $O_s(n)$, $N = mn$. Let us assume the opposite: the number of non-synchronized rotators in a finite interval is $k < n$. Let us remove all of the cluster forming oscillators from the chain (we cut the circuit) by leaving the penultimate cluster forming oscillator as well as the aforementioned segment. Then we renumber the remaining oscillators from 1 to $k+n$. In this case, we obtain a scheme similar to one shown in Fig. 6.21 with the number of oscillators $n + k$. According to the condition, we have: $V_{n+k+1-i} = V_{n-k+i}$, $i = 1, 2, \ldots, k$. Let us write equations for the last and for the $(n - k + 1)$ synchronized oscillators: $V_{n+k-1} - V_{n+k} = RI$, $V_{n-k} - 2V_{n-k+1} + V_{n-k+2} = RI$, $V_{n+k} = V_{n-k+1}$, $V_{n+k-1} = V_{n-k+2}$. From these equations and condition $V_{n+k} = V_{n-k+1}$, $V_{n+k-1} = V_{n-k+2}$, we obtain $V_{n-k} = V_{n-k+1}$. From here, it follows that if $k \neq n$, then the system of the first n rotators does not represent a cluster forming oscillator, which is a contradiction, while if $k = n$, then $V_0 = V_1$, which represents a boundary condition.

Eventually, the final conclusion of this part of the proof is: if $\mathbf{X}_n = \mathbf{X}_{n+1}$ and if the first n oscillators are not synchronized, then the cluster structure is formed based on synchronization of cluster forming oscillator of the type $O_s(n)$. Considering this equality in system (6.13), we find that it represents Eq. (6.11) in the expanded form. In contrast, by converging (6.13), we obtain Eq. (6.11) with cluster matrix $\mathbf{B}_s$.

(2) Coming back to the beginning of the proof, we now assume that the $(n + 1)$ oscillator is synchronized with the $(n - 1)$ one, i.e. $V_{n-1} = V_{n+1}$. Writing the Kirchhoff equations for these oscillators, we obtain: $V_{n-2} - 2V_{n-1} + V_n = IR$, $V_{n+2} - 2V_{n+1} + V_n = IR$. Therefore, it follows that $V_{n-2} = V_{n+2}$. Similarly, from this and all subsequent equations, we obtain the general expression $V_i = V_{2n-i}$, $i = 1, 2, \ldots, n - 1$. Therefore, the conclusion is: in this case, the dynamical structure of the system of the $(2n - 1)$ oscillators represents a cluster structure of a simple cell (see Fig. 6.22). Taking into account condition $\mathbf{X}_{n-1} = \mathbf{X}_{n+1}$ in system (6.13), we find that it represents an extended form of the equation of cluster forming oscillator $O_a(n)$ (6.12) with cluster matrix $\mathbf{B}_a$. The proof that, in this case, the complete circuit of the cluster structure of the chain consists of an integer number of simple cell's circuits (see Fig. 6.24) is carried out analogously to one considered in the first section of the proof.

The final conclusion is: $O_s(n)$ and $O_a(n)$ constitute a full set of types of structure forming objects that cover all types of possible cluster structures that could be realized in a chain of oscillators.

Remark In the proof of the theorem, the condition of chain homogeneity was not used. We have proceeded only from the condition that a certain structure in the chain is realized. This means that the theorem also holds for an inhomogeneous chain. It would have followed from the proof that this heterogeneity should have a periodically ordered form; that is, to implement a particular structure in the chain, an individual order of non-identical oscillators should be organized in accordance with the circuit of the given cluster structure.

Conclusion 1. On a number of structures in a chain for a given number N. If p is the number of all even products (simple and composite) of number N, excluding the number itself and 1, then in a chain consisting of N elementary oscillators, there exist p cluster structures based on $O_s(n)$ for different n. If q is the number of all odd multipliers (simple and composite) of number N, including this number if it is odd, then there exist q cluster structures based on a simple cell of cluster forming oscillator $O_a(n)$ for different n. The full number of cluster structures in a chain with a given number N equals $p + q$.

Proof First, according to the theorem, all cluster structures in the chain are synthesized based on $O_s(n)$ or based on a simple cell of the type $O_a(n)$. Second, all cluster structures consist of an integer number of corresponding blocks (see Figs. 6.21, 6.24). For a cluster structure based on $O_s(n)$, the following equality is valid: $N = mn$, $m \geq 2, n \geq 2$. Hence, we obtain the number of cluster structures on its base that is equal to p. For structures based on SC, $O_a(n)N = m(2n - 1)$, $m \geq 1, n \geq 2$, and that means that the number of such structures equals q. The total number of structures equals $p + q$. For an inhomogeneous chain, this means that such and only such a number of structures can be produced in a chain by manipulating its heterogeneity.

Example. Consider $N = 100$; in this case, $p = 7$, and $q = 2$, i.e. 9 structures can be realized in the chain. Let us list these structures: $O_s(2)$ (the structure contains $m = 50$ pieces of such cluster forming oscillators), $O_s(4)$ (contains 25 pieces), $O_s(5)$ (contains 20 pieces), $O_s(10)$ (contains 10 pieces), $O_s(20)$ (contains 5 pieces), $O_s(25)$ (contains 4 pieces), $O_s(50)$ (contains 2 pieces), $O_a(13)$ (contains $m = 4$ simple cells $O_a(13)$), $O_a(3)$ (contains 20 pieces).

Conclusion 2. On integral manifolds of a chain of oscillators as a dynamical system. Based on the arithmetic properties of the number of oscillators in the chain N, as well as on Conclusion 1, it is possible to obtain all integral manifolds of the chain presented in Sect. 6.1. Their mathematical representations are obtained by equating the vectors of dynamic variables of synchronized oscillators. According to the theorem proven, it can be stated that there are no other "cluster" integral manifolds in the chain, other than those indicated in Sect. 6.1.

6.4 C-Oscillators of a Ring and Fusion of Cluster Structures

Along with the chain, the ring of oscillators is one of the basic models of the theory of oscillations and waves [27]. In particular, the ring is a model for studying autowave processes in active media with periodic boundary conditions, for studies of the dynamics of collective phase synchronization systems, ring systems of superconducting transitions, and it is also used in modelling biological, chemical, and other systems [3, 5, 28–32].

In the development of the theory of cluster structures, in what follows, we will consider cluster-forming objects of the ring and establish all existing types of cluster structures [33].

Cut cluster structures of the ring. Suppose that the circuit of the cluster structure of the ring contains a cut pair. Cutting this pair reduces the circuit in the ring to the equivalent circuit in the chain. This means that the basic, known-to-us cluster

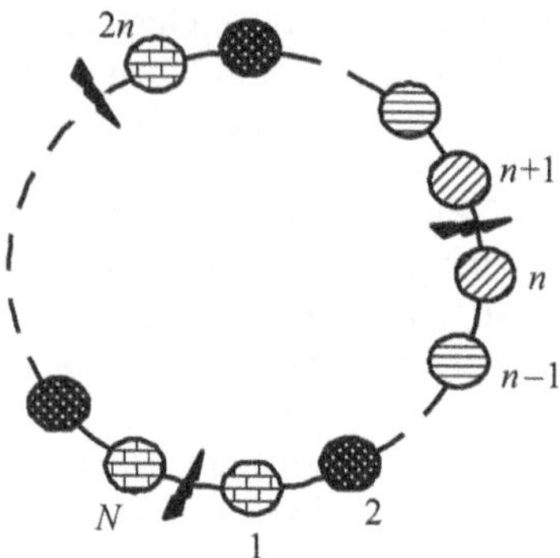

Fig. 6.37 A fusion of the cluster structure of a ring based on cluster forming oscillator of the type $O_s(n)$

forming oscillators of the chain, $O_s(n)$ and $O_a(n)$, are also the basic ones for the ring. Definitions of these cluster forming oscillators are provided in Sect. 6.3.

A fusion of the schemes of the cut cluster structures is carried out in a similar way as it was done for the chain, but with one remark: the number blocks of cluster forming oscillators $O_s(n)$ in a circuit of the cluster structure is always even. This is not difficult to explain: first, if, as a result of cutting the cut pair, the circuit of the ring turns into a circuit of the chain, then conversely, the circuit of the chain turns into a circuit of the ring as a result of connecting the oscillators. Second, since in this procedure the above oscillators are "same-coloured", the corresponding structure in the chain is centrally symmetric. Hence, the conclusion: the number of circuit blocks $O_s(n)$ in a corresponding structure is always even. Figure 6.37 shows a circuit of a cut cluster structure synthesized based on $O_s(n)$.

As an example, Fig. 6.38 shows a cut cluster structure as a synchronization of the pair of cluster forming oscillators $O_s(3)$ in a homogeneous ring of $N = 6$ diffusive coupled Chua's oscillators. The following parameter set has been used: $\{\alpha, \beta, \gamma, m_0, m_1, \varepsilon\} = \{9.5; 14; 0, 1; -1/7; 2/7; 0.34\}$. The first and sixth, as well as the third and fourth, elementary oscillators are two cut pairs of the cluster structure.

In contrast to structures based on $O_s(n)$, the number of simple cells based on $O_a(n)$ in the cluster structure of a ring can be arbitrary—even or odd. Figure 6.39 shows a fusion of the circuit of a cut cluster structure based on simple cells $O_a(n)$.

To illustrate the aforesaid matter, Fig. 6.40 shows a result of the numerical experiment: it shows a cut cluster structure as a synchronization of two simple cells $O_a(2)$ in a homogeneous ring of $N = 6$ Chua's oscillators. The following parameter set has been used: $\{\alpha, \beta, \gamma, m_0, m_1, \varepsilon\} = \{9.5; 14; 0.1; -1/7; 2/7; 0.2175\}$.

Uncut cluster structures of the ring. Suppose that the structure of the ring contains no cut pairs. In this case, it represents a simple cell and, therefore, its circuit shall be converged to the circuit of a certain (sought) cluster forming oscillator. In a homogeneous ring, it is natural to associate procedures of convergence with the different types of symmetry.

1^0. *Mirrored cluster structures of the ring.* According to our knowledge of a chain, one of the procedures of convergence—folding of a circuit of the structure of the

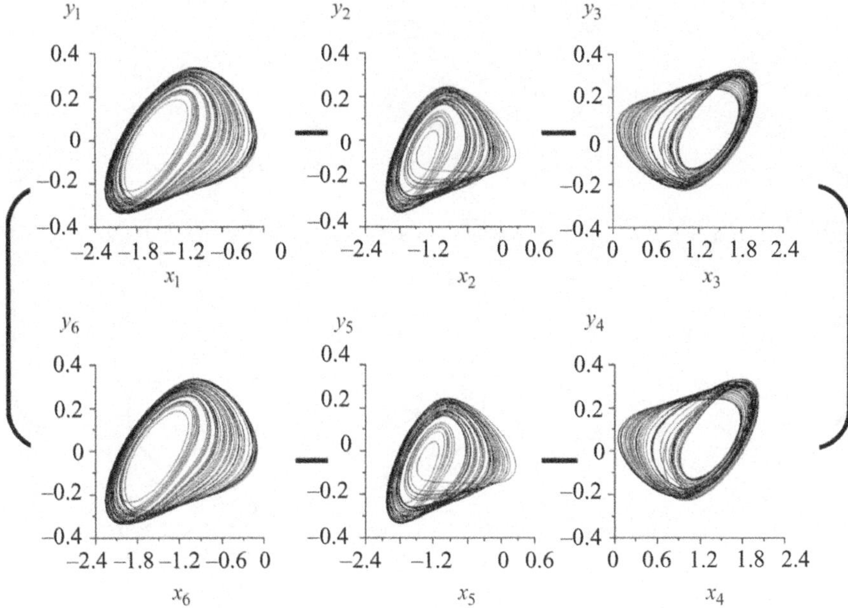

Fig. 6.38 Cut cluster structure based on a pair of cluster forming oscillators of the type $O_s(3)$

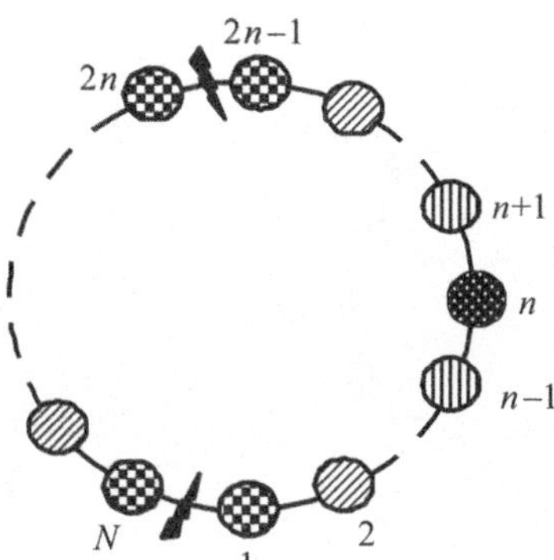

Fig. 6.39 A fusion of the cluster structure in a ring based on simple cell $O_a(n)$

simple cell along the its axis of symmetry, can be also considered as a mirroring. In the case of a ring, this axis is a diameter that passes through the centers of pictures of the oscillators (we consider the ring as a circumference with a homogeneous distribution of oscillators). For another case of the axis of symmetry—diameter that passes through centers of couplings, the structure would be a cut structure, which has been already discussed above.

Suppose that the circuit of the cluster structure is symmetric with respect to a certain diameter AF. By folding the circuit over this diameter and aligning the same-coloured images, we obtain the case shown in Fig. 6.41a. Elementary oscillators are

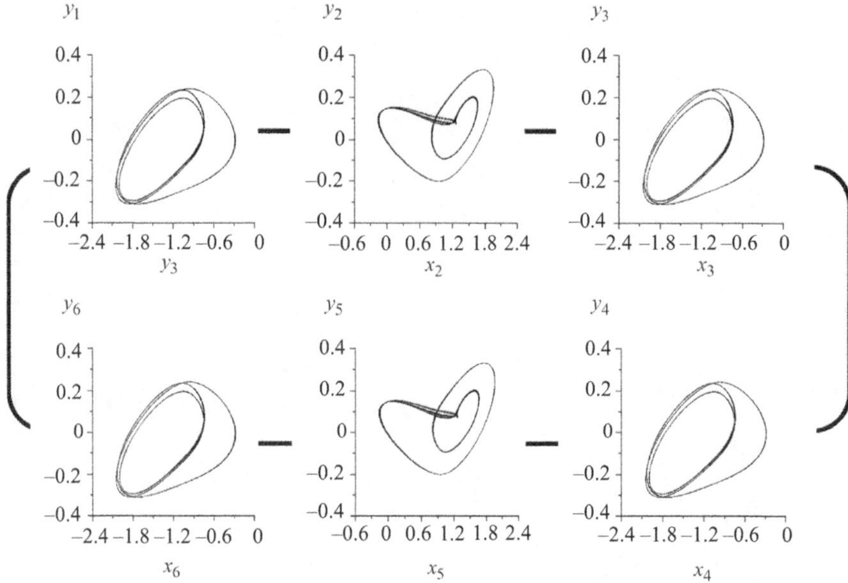

Fig. 6.40 Cut cluster structure as a synchronization of a pair of simple cells of the type $O_a(2)$

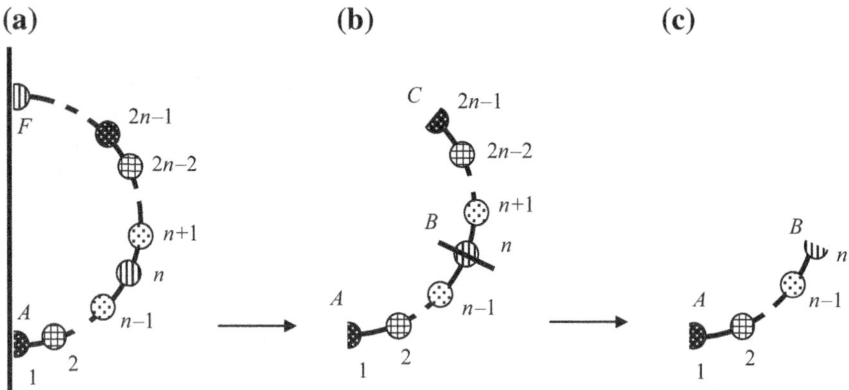

Fig. 6.41 Convergence of the circuit of a "mirrored" cluster structure of the simple cell

placed over a radial arc and the first and last ones are half-oscillators. The physical sense of these half-oscillators is clear. If all oscillators of the arc are not synchronized (have different colours), then their circuit is the sought cluster forming oscillator and the procedure of convergence is finished.

Let us, however, suppose that there is a general case when there are synchronized oscillators among the oscillators of the arc (same-colored ones). Then the circuit of oscillators of the arc AF represents an integer number of circuits of cluster forming oscillators. The circuit of a basic cluster forming oscillator shall be such that it would occupy the whole arc AF (with half-oscillators at the ends). In the opposite case, the transition to this case would be absent for the increased number of oscillators of the cluster forming oscillator.

Staring from oscillator F, we select all successive "multicolored" elementary oscillators—an arc of oscillators. Continuing to bend this and similar arcs through the middle of the final oscillator, we obtain the picture shown in Fig. 6.41c. Figure 6.41b shows the arc before the last iteration. Elementary oscillators on the AF arc represent the circuit of the sought cluster forming oscillator, the basic one for the cluster structure, which we will call the mirror cluster structure. We denote it as $O_a^r(n)$.

Definition 6.3 A system of $n \geq 2$ coupled elementary oscillators of the form

$$\dot{\mathbf{X}} = \mathbf{G}(\mathbf{X}) - \varepsilon \mathbf{B}_a^r \otimes \mathbf{C}\mathbf{X}, \tag{6.14}$$

$$\mathbf{X} = (\mathbf{X}_1, \mathbf{X}_2, \ldots, \mathbf{X}_n)^T, \quad \mathbf{G}(\mathbf{X}) = (\mathbf{F}_1(\mathbf{X}_1), \mathbf{F}_2(\mathbf{X}_2), \ldots, \mathbf{F}_n(\mathbf{X}_n))^T,$$

$$\mathbf{B}_a^r = \begin{pmatrix}
\begin{matrix} 2 & -2 & 0 & 0 \\ -1 & 2 & -1 & 0 \\ 0 & -1 & 2 & -1 \\ 0 & 0 & -1 & 2 \end{matrix} & & & 0 \\
& \ddots & & \\
& & & \begin{matrix} 2 & -1 & 0 & 0 \\ -1 & 2 & -1 & 0 \\ 0 & -1 & 2 & -1 \\ 0 & 0 & -2 & 2 \end{matrix} \\
\end{pmatrix}$$

Is called the mirror cluster oscillator $O_a^r(n)$ and conditioned that, in the phase space, there exists attractor $A_a^r(n)$, such that for $\forall (X_1, X_2, \ldots, X_n) \in A_a^r(n), X_i \neq X_j, i \neq j, i = \overline{1, n}, j = \overline{1, n}$.

Equation (6.14) can be obtained in different ways, and in particular, from Eq. (6.13), for which the following boundary conditions should be taken: $\mathbf{X}_2 \equiv \mathbf{X}_0$, $\mathbf{X}_{n+1} \equiv \mathbf{X}_{n-1}$.

Figure 6.42 shows a mirror cluster structure in a homogeneous ring of $N = 6$ Chua's oscillators. The following parameter set has been used: $\{\alpha, \beta, \gamma, m_0, m_1, \varepsilon\} = \{9.5; 14; 0.1; -1/7; 2/7; 0.25\}$. The surface of a "mirror" passes through the first and fourth oscillators.

The fusion of the circuit of the mirror structure is carried out in the order opposite of convergence: by a series of a chain of mirror reflections of the circuit of "mirror" cluster forming oscillator: its mirror reflection is attached to the arc of the oscillators AB and so on, up to the closure of the circuit of the structure into the ring.

It is not difficult to derive a formula for coupling the number of oscillators in the ring with the number of cluster forming oscillators in the fusion of the structure

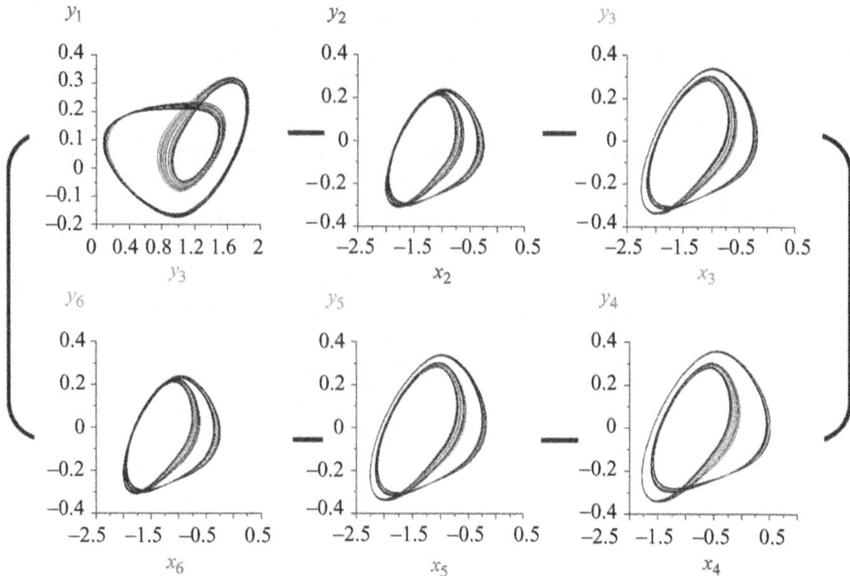

Fig. 6.42 Mirror cluster structure based on $O_a^r(4)$

and the number of elementary oscillators in the cluster forming oscillator itself: $N = 2m(n-1)$, where m is the number of cluster forming oscillators that "cover" a half of the ring. From this formula, it follows that if r is the number of all multipliers of number $N/2$, including the number itself, then in the ring there exist r mirror cluster structures based on cluster forming oscillator $O_a^r(n)$. For instance, for $N = 12$, there exist three structures based on $O_a^r(n)$: one consisting of two cluster forming oscillators $O_a^r(7)$, one consisting of four cluster forming oscillators $O_a^r(4)$, and one consisting of six cluster forming oscillators $O_a^r(3)$. For $N = 6$, there exists one mirror structure in the ring based on the synchronization of $O_a^r(4)$, the example of which is shown in Fig. 6.42.

2^0. *Cyclic cluster structures of the ring.* This type of cluster structure is associated with the symmetry of the rotation of the ring. However, we come to the existence of such structures and define their base cluster forming oscillator based on the physical principles reflected in the elementary transformations and the coagulability of circuits of the structures.

Suppose that a certain uncut n-cluster structure is realized in the ring. We reduce all of the elementary oscillators of one cluster to a single node (by short circuiting, see Fig. 6.43a). As a result, we obtain a figure resembling an m-petal "rose" (see Fig. 6.43b). Suppose that, among the oscillators of each petal, there are no single-coloured ones (from different clusters). Now we prove that, in this case, the order of the oscillators on the petals is the same: continue the convergence of the circuit of the cluster structure by turning the "petals" until the matching of same-coloured oscillators (short circuits). As a result, a ring of unsynchronized oscillators is formed

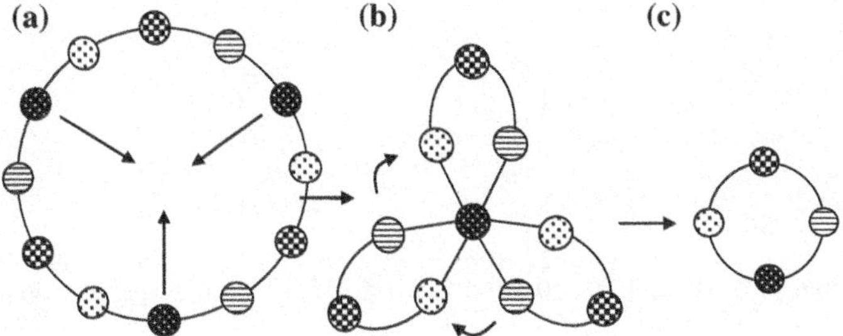

Fig. 6.43 Convergence of a circuit of the cyclic cluster structure

(see Fig. 6.43c). This ring also represents the sought cluster forming oscillator of "cyclic" structures.

The fusion of the circuit of the cyclic structure is carried out in the reverse order: we assume that there are m identical rings (cluster forming oscillators) superimposed on each other. Further, this system unfolds with respect to the image of any oscillator (more precisely, a node of m identical oscillators) into an m-petal "rose". Then the images of the node's oscillators blossom in their places. The same result is obtained as a result of successive rotations of the arc of "multi-coloured" oscillators.

We denote the basic cluster forming oscillator of cyclic cluster structures as $O_s^r(n)$.

Definition 6.4 A system of $n \geq 3$ coupled elementary oscillators of the form

$$\dot{\mathbf{X}} = \mathbf{G}(\mathbf{X}) - \varepsilon \mathbf{B}_s^r \otimes \mathbf{CX}, \tag{6.15}$$

$$\mathbf{X} = (\mathbf{X}_1, \mathbf{X}_2, \ldots, \mathbf{X}_n)^T, \, \mathbf{G}(\mathbf{X}) = (\mathbf{F}_1(\mathbf{X}_1), \mathbf{F}_2(\mathbf{X}_2), \ldots, \mathbf{F}_n(\mathbf{X}_n))^T,$$

$$\mathbf{B}_s^r = \begin{pmatrix}
2 & -1 & 0 & 0 & \ldots & 0 & 0 & 0 & -1 \\
-1 & 2 & -1 & 0 & \ldots & 0 & 0 & 0 & 0 \\
0 & -1 & 2 & -1 & \ldots & 0 & 0 & 0 & 0 \\
0 & 0 & -1 & 2 & \ldots & 0 & 0 & 0 & 0 \\
\vdots & \vdots & \vdots & \vdots & \ddots & \vdots & \vdots & \vdots & \vdots \\
0 & 0 & 0 & 0 & \ldots & 2 & -1 & 0 & 0 \\
0 & 0 & 0 & 0 & \ldots & -1 & 2 & -1 & 0 \\
0 & 0 & 0 & 0 & \ldots & 0 & -1 & 2 & -1 \\
-1 & 0 & 0 & 0 & \ldots & 0 & 0 & -1 & 2
\end{pmatrix}$$

Is called the cyclic cluster forming oscillator $O_s^r(n)$ and is conditioned that, in the phase space, there exists attractor $A_s^r(n)$ such that for $\forall (\mathbf{X}_1, \mathbf{X}_2, \ldots, \mathbf{X}_n) \in A_s^r(n)$, $\mathbf{X}_i \neq \mathbf{X}_j, i \neq j, i = \overline{1, n}, j = \overline{1, n}$.

It is not difficult to see that Eq. (6.15) represents a ring of n oscillators. Figure 6.44 shows a cyclic cluster structure obtained in a homogeneous ring of $N = 6$ Chua's oscillators. The following parameter set has been used: $\{\alpha, \beta, \gamma, m_0, m_1, \varepsilon\} =$

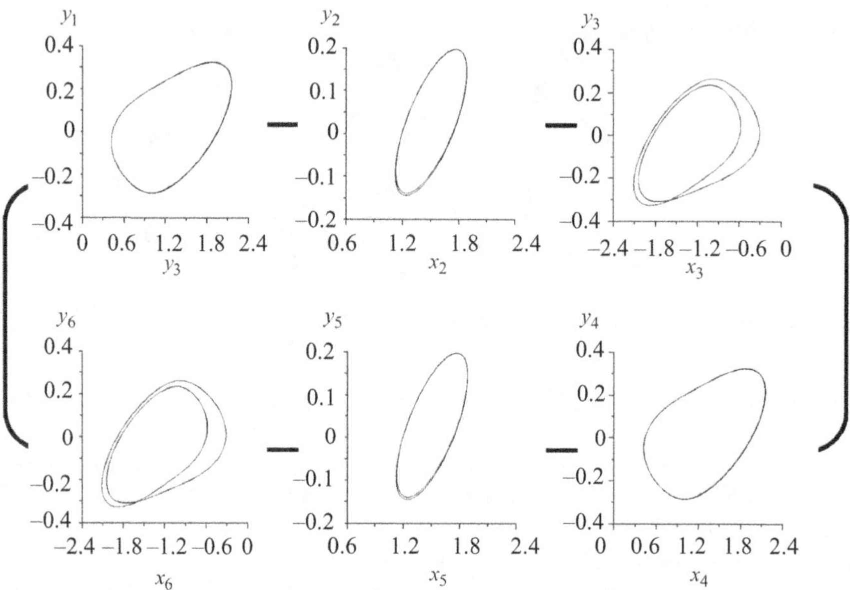

Fig. 6.44 Cyclic cluster structure based on $O_s^r(3)$

$\{9.5;\ 14;\ 0.1;\ -1/7;\ 2/7;\ 0.08\}$. It is evident that the number of synchronized cluster forming oscillators and the number of clusters are related to the number of oscillators in the ring by a simple relation: $N = mn$. This allows formulation of a simple statement: if s is the number of all multipliers of number N, excluding N itself, then the number of cyclic structures in the ring equals s. For instance, for $N = 12$, there exist four cyclic structures based on $O_s^r(n)$: a structure of two cluster forming oscillators of the type $O_s^r(6)$, a structure of three cluster forming oscillators of the type $O_s^r(4)$, a structure of four cluster forming oscillators of the type $O_s^r(3)$, and a structure of six cluster forming oscillators of the type $O_s^r(2)$. For $N = 6$ there exist two cyclic cluster structures, one of which (three-cluster one) is shown in Fig. 6.44.

Remark Cluster structures synthesized in a ring based on a mirror cluster forming oscillator with two elementary oscillators, along with an axial one, also have a cyclic symmetry (see Fig. 6.45).

The duality in the question of which type this structure belongs to is removed if we write the equations of the ring in the form of a system of coupled cluster forming oscillators. In this case, we will find that the structure is formed on the basis of a mirror cluster forming oscillator.

Above, we have defined four types of cluster structures of the ring. The inner conviction suggests (everything is taken into account, nothing else can be) that a complete group of cluster forming oscillators is found. Let us try to prove it. We formulate a simple lemma.

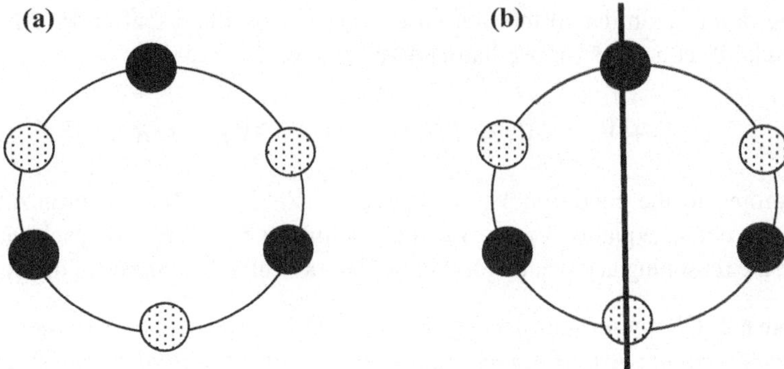

Fig. 6.45 Structure with double symmetry

Lemma 6.1

(1) *If the oscillators at the ends of two parallel chords are synchronized in pairs
     (see Fig. 6.46a), then the oscillators at the ends of all of the chords of the bundle
     are synchronized in pairs.*

(2) *If the oscillators at the ends of two intersecting chords are synchronized in
     pairs (see Fig. 6.46b), then the oscillators at the ends of all of the chords of the
     bundle are synchronized in pairs (oscillators are distributed along the ring in
     an appropriate way).*

Proof

(1) Consider oscillators of chord *EF*. We write the Kirchhoff equations for
     oscillators *A* and *B*:

$$V_C + V_E - 2V_A = I_A R, \; V_D + V_F - 2V_B = I_B R.$$

According to condition $V_A = V_B$, $V_C = V_D$, $I_A = I_B$ (synchronization). Hence,
we obtain the equality $V_E = V_F$, i.e. oscillators *E* and *F* are synchronized. Further,
the reasoning is repeated for the oscillators of all other chords of the bundle.

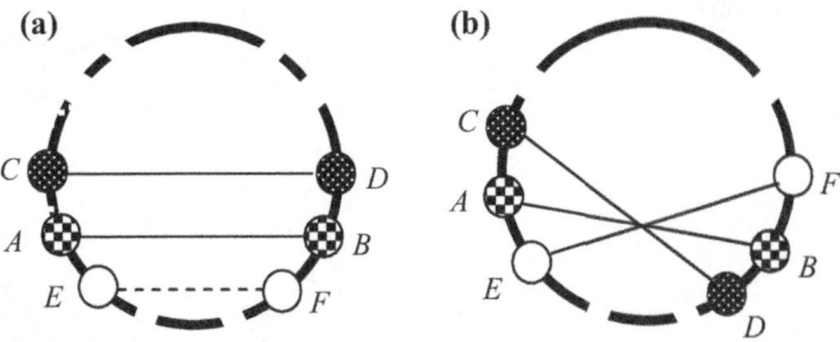

Fig. 6.46 Example for Lemma 6.1

(2) The proof is similar to the first case. Consider oscillators of chord EF. The
Kirchhoff equations for oscillators A and B have the form:

$$V_C + V_E - 2V_A = I_A R, \quad V_D + V_F - 2V_B = I_B R.$$

According to the condition $V_A = V_B, V_C = V_D, I_A = I_B$ (synchronization).
Hence, we obtain equality $V_E = V_F$, i.e. oscillators E and F are synchronized.
Further, the reasoning is repeated for the oscillators of all other chords of the bundle.

Theorem 6.2 *Cluster forming oscillators $O_s(n)$, $O_a(n)$, $O_a^r(n)$ and $O_s^r(n)$ constitute
a full set of types of cluster forming oscillators in a ring of oscillators and define all
types of cluster structures that can exist in the ring.*

Proof Let us assume that some cluster structure is realized in the ring. We select
the arc AF with the maximum number n of unsynchronized oscillators and number
them counter-clockwise, starting at one end (see Fig. 6.47).

By the condition, the oscillator G is synchronized with one of the oscillators of
the arc AF. We will consider various cases. If in each of them the cluster structure is
represented by one of the four of these cluster forming oscillators, then this will be
a proof of the theorem, excluding, of course, physically impossible cases.

(1) Suppose that oscillator G is synchronized with oscillator F. In this case, this
pair of oscillators is a cut pair and, therefore, this cluster structure is also a cut
structure. The basic cluster forming oscillators of such structure are $O_s(n)$ and
$O_a(n)$.
(2) Suppose that oscillator G is synchronized with oscillator E. According to Lemma
1 (a special case when one of the chords is tied to the point), oscillators at the
ends of all chords that are parallel to GE are synchronized, and we have a
structure that is symmetric with respect to the diameter passing through the
oscillator F. If the number of oscillators in the ring is odd, then the last short
chord of the bundle contracts the cut pair; that is, there is again a cut structure
with the basic cluster forming oscillator $O_a(n)$. If the number of oscillators is

Fig. 6.47 For Theorem 6.2

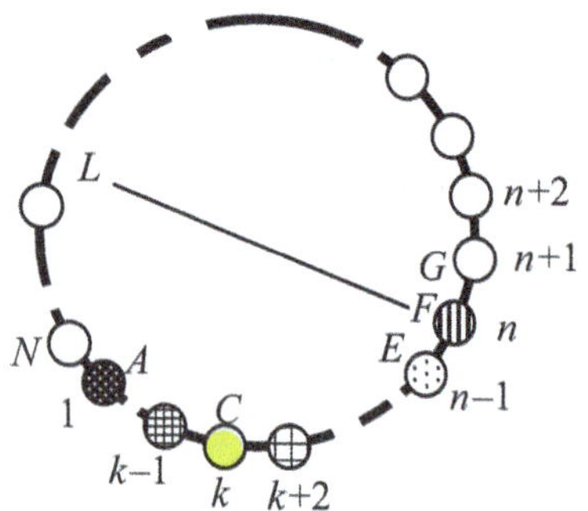

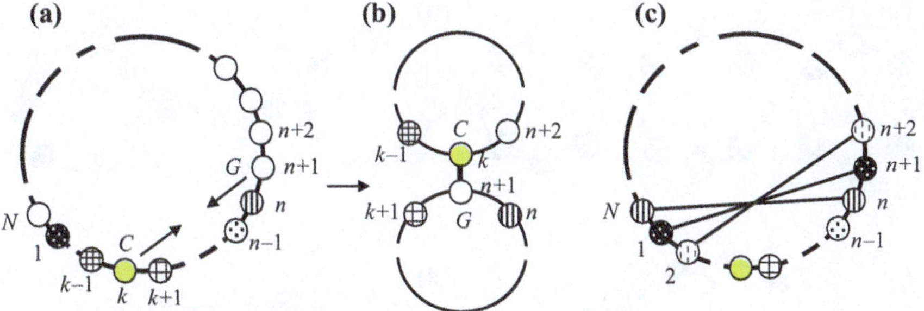

Fig. 6.48 Equivalent transformations of the circuit of a cluster structure

even, then diameter FL passes through the middle of the oscillators. Suppose that, among the oscillators of arc LG, there is no cut pair [otherwise the structure is cut and its cluster forming oscillator is $O_a(n)$]. In such a case, this cluster structure (symmetric with respect to diameter FL) is a "mirror" and its cluster forming oscillator is $O_a^r(n)$.

(3) Suppose that oscillator G is synchronized with oscillator C. We make equivalent transformations of the cluster structure: synchronized oscillators C and G are "tightened" into a node, and then we "move" them along different circles (see Fig. 6.48a, b).

We write the Kirchhoff equations for oscillators G and C in the first and second positions of the system:

$$V_{n+2} + V_n - 2V_G = I_G R\,, \quad V_{k-1} + V_{k+2} - 2V_C = I_C R\,,$$
$$V_{k-1} + V_{n+2} - 2V_C = I_C R\,, \quad V_{k+1} + V_n - 2V_G = I_G R\,.$$

Taking into account equalities $V_G = V_C$, $I_G = I_C$ (synchronization), we obtain $V_n = V_{k-1}$ and $V_{n+2} = V_{k+1}$. For any $k > 1$, these equations contradict the condition that the first n oscillators are not synchronized, and when $k = 1$, taking into account that $V_0 = V_N$, we obtain equalities $V_n = V_N, V_{n+2} = V_2$. According to Lemma 6.1, the latter equations define a cyclic cluster structure with basic cluster forming oscillator $O_s^r(n)$ (see Fig. 6.48c). The theorem is proven.

Conclusion. If p is the number of all multipliers (simple and composite) of number $N/2$, including the number itself, and q is the number of all odd multipliers of number N, including the number itself if it is odd, then the full number of cut structures in the ring equals $p+q$. If r is the number of all multipliers of number $N/2$, including the number itself, then in the ring there exist r "mirror" cluster structures based on cluster forming oscillator $O_a^r(n)$. If s is the number of all multipliers N, excluding N, then the number of cyclic structures based on $O_s^r(n)$ equals s. The full number of cluster structures of all types equals $p + q + r + s$.

The proof of the conclusion is actually given above, during the investigation of types of cluster oscillators.

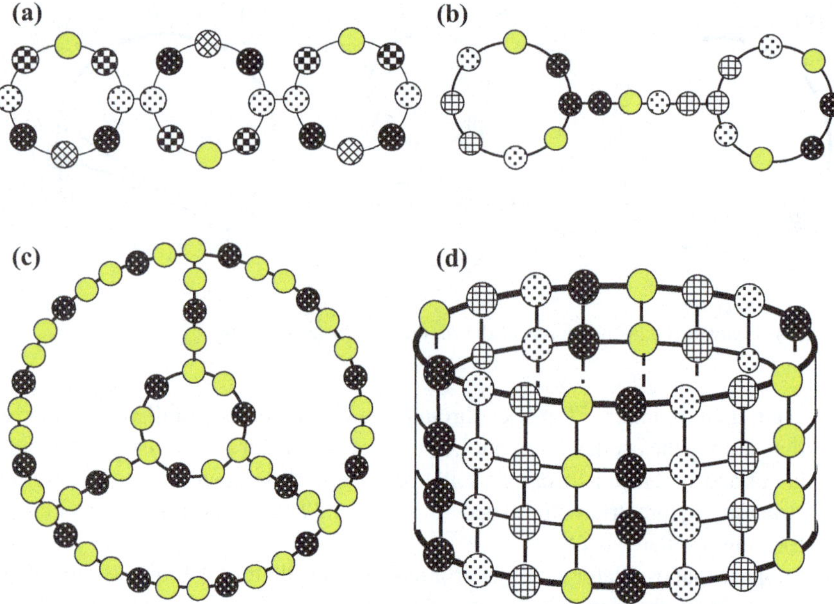

Fig. 6.49 Examples of cluster structures based on different cluster structures and based on the different cluster *forming* oscillators of the ring: **a** structure based on cluster forming oscillator $O_a^r(5)$, **b** structure based on cluster forming oscillator $O_s(4)$, **c** structure based on cluster forming oscillator $O_a(2)$, **d** structure based on cluster forming oscillator $O_s^r(4)$

For example, if $N = 12$, then, according to *Conclusion*, there are 11 cluster structures in the ring. Listing them is not difficult. Note that in the case of a simple N, there exists a unique simple cell (it is also a cut one) in the ring based on simple cell $O_a(n)$ with number of clusters $n = (N + 1)/2$.

At the end of this section, we also note that the above types of cluster forming oscillators can serve as the basis for the circuits of structures, not only in the ring, but also in other lattices, for which the ring is an element of their construction (see Fig. 6.49).

6.5 Stability of Cluster Structures

The phase space of homogeneous lattices of oscillators is complicated. In the general case, along with an attractor representing the cluster structure of interest (for example, $A_s(n)$, $A_a(n)$, etc.), there also exist other attractors in the phase space. These attractors can include attractor $A(1)$, which corresponds to spatially-homogeneous state of the lattice, which occurs always. This circumstance substantially complicates the problem of studying the stability of structures, including narrowing the set of research methods (excluding the second Lyapunov method). In addition, if the intrinsic properties of the main attractor $A(1)$, which we dealt with earlier, do not depend on parameter of the lattice ε, then this cannot be said of cluster attractors. Their properties, in particular the Lyapunov exponents, as well as the existence of these attractors, are determined by this parameter, which excludes the possibility of formulating the stability conditions explicitly, like in the case of spatially homogeneous structures. This creates additional difficulties in the refinement of the analytical conditions of stability and suggests additional numerical or full-scale studies.

Further, we assume that attractors of the cluster structures of interest exist and their properties are known. In particular, we assume that the maximal Lyapunov exponents of attractors are known.

(1) *Stability of the simplest cluster structure based on $O_s(n)$ during the parallel fusion.*

Figure 6.50 shows a circuit of cluster forming oscillator $O_s(n)$ as well as a circuit of a cluster structure on its base.

Equations of the lattice are composed using the experience from the previous sections. For vectors of the dynamical state of C-oscillator $\mathbf{Y}_j = \mathrm{col}(\mathbf{X}_{1j}, \mathbf{X}_{2j}, \dots, \mathbf{X}_{nj})$, $j = \overline{1, N_s}$, these equations take the form

$$\dot{\mathbf{Y}}_j = \mathbf{G}(\mathbf{Y}_j) - \varepsilon(\mathbf{I}_n \otimes \mathbf{C})(-\mathbf{Y}_{j-1} + 2\mathbf{Y}_j - \mathbf{Y}_{j+1}),$$
$$j = 1, 2, \dots N. \tag{6.16}$$

With boundary conditions: $\mathbf{Y}_0 = \mathbf{Y}_1$, $\mathbf{Y}_{N_s} = \mathbf{Y}_{N_s+1}$.

Let us study the local stability of attractor $A_s(n)$, by linearizing (6.16) in the vicinity of solution $\mathbf{Y}_1 = \mathbf{Y}_2 = \dots = \mathbf{Y}_{N_s} = \mathbf{x}(t) \in A_s(n)$. With respect to vector $\mathbf{U} = (\mathbf{U}_1, \mathbf{U}_2, \dots, \mathbf{U}_{N_s-1})^T$, where $\mathbf{U}_i = \mathbf{Y}_i - \mathbf{Y}_{i+1}$, the corresponding linear system, written in a form of one equation, has the form

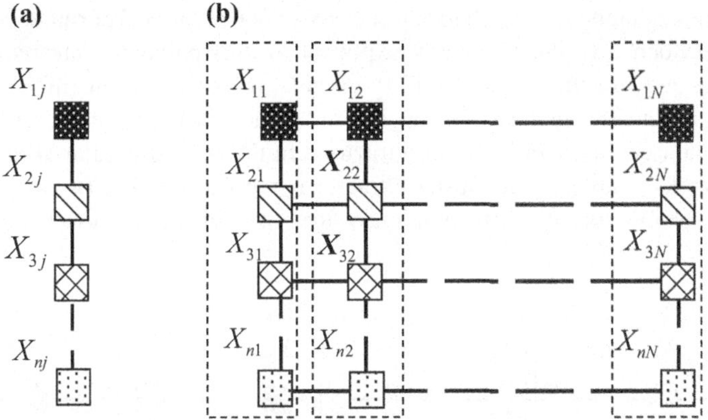

Fig. 6.50 A circuit of C-oscillator $O_s(n)$ (**a**); and tape cluster structure on its basis, (**b**)

$$\dot{\mathbf{U}} = \left(\mathbf{I}_p \otimes \mathbf{J}(\xi) - \varepsilon \mathbf{D}_p \otimes \mathbf{I}_n \otimes \mathbf{C}\right)\mathbf{U}, \tag{6.17}$$

where $p = N_s - 1$, $\mathbf{J}(\xi) = \mathrm{diag}(\mathbf{J}_1(\xi_1), \mathbf{J}_2(\xi_2), \ldots, \mathbf{J}_n(\xi_n)) - \varepsilon \mathbf{B}_n \otimes \mathbf{C}$ is the Jacobi matrix of cluster forming oscillator $O_s(n)$, $\mathbf{J}_k(\xi_k)$ is the Jacobi matrix of the elementary oscillator, and $\mathbf{D}_p$ is a matrix of the form

$$\mathbf{D}_p = \begin{pmatrix}
2 & -1 & 0 & 0 & & & & & & \\
-1 & 2 & -1 & 0 & & & 0 & & & \\
0 & -1 & 2 & -1 & & & & & & \\
0 & 0 & -1 & 2 & & & & & & \\
& & & & \ddots & & & & & \\
& & & & & & 2 & -1 & 0 & 0 \\
& & 0 & & & & -1 & 2 & -1 & 0 \\
& & & & & & 0 & -1 & 2 & -1 \\
& & & & & & 0 & 0 & -1 & 2
\end{pmatrix}.$$

Equation (6.17) complies with Lemma 5.1 and, consequently, with Theorem 5.1 under the condition of replacement of matrix $\mathbf{C}$ (in Lemma and in Theorem) by matrix $\mathbf{I}_n \otimes \mathbf{C}$. Taking this into account and considering that eigenvalues of matrix $\mathbf{D}_p$ are known to us, we obtain conditions of stability of the cluster structure in a form of the following inequalities: $\varepsilon > \lambda_s(n)\big/4\sin^2\pi\big/2N_s$, if $\mathbf{C} = \mathbf{I}_m$, and $\varepsilon > \varepsilon^*(\lambda_s(n))\big/4\sin^2\pi\big/2N_s$ in a general case of matrix $\mathbf{C}$.

As we see, the inequalities obtained for the stability of a given cluster structure and the corresponding inequalities in the case of a spatially homogeneous structure in the chain are identical in their forms: the circuit shown in Fig. 6.51 can be interpreted as a "chain of chains", so the number of elementary oscillators is replaced by the number of blocks — C-oscillators. The principal difference is that, in this case, the Lyapunov exponent depends on number ε: $\lambda_s = \lambda_s(n, \varepsilon)$. That is, these parameters are implicit in the coupling parameter, which means that additional numerical or full-scale studies of the dependence of the Lyapunov exponent on this parameter are required for the completeness of the picture. In [24], circuits, mathematical justification, and a procedure for measuring the exponents in a full-scale experiment are described. This concerns the case of chaotic cluster attractors. In the case of regular attractors, the situation looks simple: if the cluster attractor is regular, i.e. if $\lambda_s(n, \varepsilon) \leq 0$, then structure is stable for any corresponding values of ε, including the arbitrary, small ones.

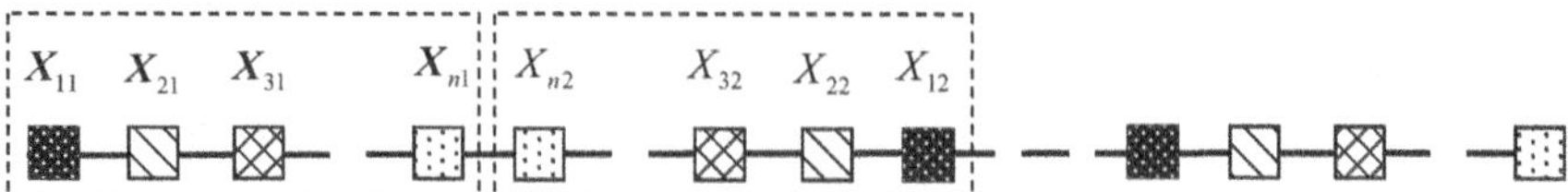

Fig. 6.51 Cluster structure based on $O_s(n)$

(2) *Stability of cluster structure based on $O_s(n)$ during the sequential fusion.*

For the beginning, let us study the case of two C-oscillators (see Fig. 6.51).

With respect to vectors $\mathbf{Y}_j = \mathrm{col}(\mathbf{X}_{1j}, \mathbf{X}_{2j}, \ldots, \mathbf{X}_{nj})$, $j = 1,\ 2$, the dynamical system has the form

$$\dot{\mathbf{Y}}_1 = \mathbf{G}(\mathbf{Y}_1) - \varepsilon(\mathbf{C}_* \otimes \mathbf{C})(\mathbf{Y}_1 - \mathbf{Y}_2),$$
$$\dot{\mathbf{Y}}_2 = \mathbf{G}(\mathbf{Y}_2) + \varepsilon(\mathbf{C}_* \otimes \mathbf{C})(\mathbf{Y}_1 - \mathbf{Y}_2), \tag{6.18}$$

where $\mathbf{C}_* = \mathrm{diag}(0, 0, \ldots, 0, 1)$ is the fusion matrix. A linearization of (6.18) by variable $\mathbf{U} = \mathbf{Y}_1 - \mathbf{Y}_2$ leads to the equation

$$\dot{\mathbf{U}} = (\mathbf{J}(\mathbf{x}) - \varepsilon 2\mathbf{C}_* \otimes \mathbf{C})\mathbf{U}. \tag{6.19}$$

Directly from Eq. (6.19), we find that a structure with regular internal dynamics is stable. This follows from the fact that matrix $\varepsilon 2\mathbf{C}_* \otimes \mathbf{C}$ is diagonal and can be considered as a perturbation of Jacobian $\mathbf{J}(\mathbf{x})$. Such perturbations do not increase the Lyapunov exponents of solutions [34, 35]. In contrast to the parallel connection, where the fusion matrix was a unit matrix ($\mathbf{C}_* \to \mathbf{I}_n$), in this case, it is impossible to formulate the stability conditions directly from Eq. (6.19) or, at least, it is difficult. To overcome technical difficulties, we formulate an additional lemma.

Lemma 6.2 *Assume that the norm of variable matrix $A(t)$ is bounded: $\|A(t)\| = \mu$, and $\lambda_{\min} > 0$ is the least root of constant symmetric matrix B. Then, for $\varepsilon > \mu/\lambda_{\min}$, a trivial solution of equation*

$$\dot{\mathbf{y}} = (\mathbf{A}(t) - \varepsilon\mathbf{B})\mathbf{y} \tag{6.20}$$

Is asymptotically stable.

Proof Suppose that $\mathbf{Y}(t,\ t_0)$ is a matriciant of equation $\dot{\mathbf{y}} = -\varepsilon\mathbf{B}\mathbf{y}$. Then $\mathbf{Y}(t,\ t_0) = e^{-\varepsilon B(t-t_0)}$ and $\|\mathbf{Y}(t,\ t_0)\| = e^{-\varepsilon\lambda_{\min}(t-t_0)}$. Taking expression $\mathbf{A}(t)\mathbf{y}$ as an external perturbation, we rewrite (6.20) in an integral form:

$$\mathbf{y}(t) = \int_{t_0}^{t} \mathbf{Y}(t, \tau)\mathbf{A}(\tau)\mathbf{y}(\tau)d\tau + \mathbf{Y}(t,\ t_0)\mathbf{c},$$

where $\mathbf{c}$ is an arbitrary constant vector. Estimating the norm of the right- and left-hand sides of this equation, we obtain the inequality

$$\|\mathbf{y}(t)\| \le \int_{t_0}^{t} \|\mathbf{Y}(t, \tau)\|\,\|\mathbf{A}(\tau)\|\,\|\mathbf{y}(t)\|d\tau + \|\mathbf{Y}(t,\ t_0)\|\,\|\mathbf{c}\|,$$

which, taking into account the proposed conditions, is transformed to the form

$$\|\mathbf{y}(t)\|e^{\varepsilon\lambda_{\min}t} \le \int_{t_0}^{t} \mu\|\mathbf{y}(\tau)\|e^{\varepsilon\lambda_{\min}\tau}d\tau + Ce^{\varepsilon\lambda_{\min}t_0},$$

where C is an arbitrary constant.

After applying the well-known lemma (about the integral inequality) [36] to the given inequality, and after further minor transformations, we obtain $\|\mathbf{y}(t)\| \le Ce^{-(\varepsilon\lambda_{\min}-\mu)(t-t_0)}$. From this inequality, it follows that: if $\varepsilon > \mu/\lambda_{\min}$, then $\mathbf{y}(t) \to 0$ for $t \to \infty$. The Lemma is proven.

Let us establish relation of Eqs. (6.19) and (6.20). To do so, we select a variable part $\mathbf{A}(t) = \mathrm{diag}(\mathbf{J}_1(\xi_1),\ \mathbf{J}_2(\xi_2),\ \ldots,\ \mathbf{J}_n(\xi_n))$ (truncated Jacobian) in the Jacobian of C-oscillator $\mathbf{J}(\xi) = \mathrm{diag}(\mathbf{J}_1(\xi_1),\ \mathbf{J}_2(\xi_2),\ \ldots,\ \mathbf{J}_n(\xi_n)) - \varepsilon\mathbf{B}_n \otimes \mathbf{C}$, while cluster matrix $\mathbf{B}_n$ will be merged with matrix $2\mathbf{C}_*$:

$$\mathbf{B} = (\mathbf{B}_n + 2\mathbf{C}_*) \otimes \mathbf{C} =
\begin{pmatrix}
\begin{matrix}
1 & -1 & 0 & 0 \\
-1 & 2 & -1 & 0 \\
0 & -1 & 2 & -1 \\
0 & 0 & -1 & 2
\end{matrix} & & \mathbf{0} \\
 & \ddots & \\
& \mathbf{0} &
\begin{matrix}
2 & -1 & 0 & 0 \\
-1 & 2 & -1 & 0 \\
0 & -1 & 2 & -1 \\
0 & 0 & -1 & 3
\end{matrix}
\end{pmatrix} \otimes \mathbf{C}.$$

In this case, Eq. (6.19) takes the form of Eq. (6.20). The norm of the truncated Jacobian:

$$\|\mathbf{A}(t)\| = \max(\|\mathbf{J}_1(\xi_1)\|,\ \|\mathbf{J}_2(\xi_2)\|,\ \ldots,\ \|\mathbf{J}_n(\xi_n)\|) = \|\mathbf{J}_0(\xi_0)\| = \mu.$$

Here we denote the "partial" Jacobian of the elementary oscillator with maximal norm $\mathbf{J}_0(\xi_0)$. It is known that the minimal eigenvalue of matrix $\mathbf{B}_n + 2\mathbf{C}_*$ equals $\lambda_{\min} = 4\sin^2\frac{\pi}{4n}$ (see Appendix II). Summarizing the aforesaid matter, we formulate stability conditions for the trivial solution (6.19) (structure): the cluster structure, in the case of a pair of coupled C-oscillators, is stable if the following inequalities are satisfied: $\varepsilon > \mu/4\sin^2\frac{\pi}{4n}$, if $\mathbf{C} = \mathbf{I}_m$, and $\varepsilon > \varepsilon^*(\mu)/4\sin^2\frac{\pi}{4n}$ in a general case of matrix $\mathbf{C}$. These conditions are sufficient.

Let us consider the case of an arbitrary number of C-oscillators of the type $O_s(n)$. With respect to vectors $\mathbf{Y}_j = \mathrm{col}(\mathbf{X}_{1j},\ \mathbf{X}_{2j},\ \ldots,\ \mathbf{X}_{nj})$, $j = 1, 2, \ldots, k$, we obtain the following system of equations:

$$\dot{\mathbf{Y}}_1 = \mathbf{G}(\mathbf{Y}_1) + \varepsilon(\mathbf{C}_* \otimes \mathbf{C})(\mathbf{Y}_2 - \mathbf{Y}_1),$$

$$\dot{\mathbf{Y}}_2 = \mathbf{G}(\mathbf{Y}_2) - \varepsilon(\mathbf{C}_* \otimes \mathbf{C})(\mathbf{Y}_2 - \mathbf{Y}_1) + \varepsilon(\mathbf{C}^* \otimes \mathbf{C})(\mathbf{Y}_3 - \mathbf{Y}_2),$$

$$\dot{\mathbf{Y}}_3 = \mathbf{G}(\mathbf{Y}_3) - \varepsilon\big(\mathbf{C}^* \otimes \mathbf{C}\big)(\mathbf{Y}_3 - \mathbf{Y}_2) + \varepsilon(\mathbf{C}_* \otimes \mathbf{C})(\mathbf{Y}_4 - \mathbf{Y}_3),$$

$$\vdots$$

$$\dot{\mathbf{Y}}_k = \mathbf{G}(\mathbf{Y}_k) + \varepsilon\big(\mathbf{C}^* \otimes \mathbf{C}\big)(\mathbf{Y}_{k-1} - \mathbf{Y}_k), \tag{6.21}$$

where $\mathbf{C}_* = \mathrm{diag}(0, 0, \ldots, 0, 1)$ and $\mathbf{C}^* = \mathrm{diag}(1, 0, \ldots, 0, 0)$ are the fusion matrices.

System (6.21) is written for odd k. In the case of even ones, in the last equation the following change is to be done: $\mathbf{C}^* \to \mathbf{C}_*$. We study sufficient conditions of stability of the cluster structure.

With respect to vectors $\mathbf{U}_j = \mathbf{Y}_j - \mathbf{Y}_{j+1}$, the linearized system has the form:

$$\dot{\mathbf{U}}_1 = \mathbf{J}(\xi)\mathbf{U}_1 - \varepsilon\big(2(\mathbf{C}_* \otimes \mathbf{C})\mathbf{U}_1 - \big(\mathbf{C}^* \otimes \mathbf{C}\big)\mathbf{U}_2\big),$$

$$\dot{\mathbf{U}}_2 = \mathbf{J}(\xi)\mathbf{U}_2 - \varepsilon\big(-(\mathbf{C}_* \otimes \mathbf{C})\mathbf{U}_1 + 2\big(\mathbf{C}^* \otimes \mathbf{C}\big)\mathbf{U}_2 - (\mathbf{C}_* \otimes \mathbf{C})\mathbf{U}_3\big),$$

$$\dot{\mathbf{U}}_3 = \mathbf{J}(\xi)\mathbf{U}_3 - \varepsilon\big(-\big(\mathbf{C}^* \otimes \mathbf{C}\big)\mathbf{U}_2 + 2(\mathbf{C}_* \otimes \mathbf{C})\mathbf{U}_3 - \big(\mathbf{C}^* \otimes \mathbf{C}\big)\mathbf{U}_4\big),$$

$$\vdots$$

$$\dot{\mathbf{U}}_{k-1} = \mathbf{J}(\xi)\mathbf{U}_{k-1} - \varepsilon\big(-\big(\mathbf{C}^* \otimes \mathbf{C}\big)\mathbf{U}_{k-2} + 2(\mathbf{C}_* \otimes \mathbf{C})\mathbf{U}_{k-1}\big). \tag{6.22}$$

System (6.22), written in a form of one equation has the form

$$\dot{\mathbf{U}} = \mathbf{I}_{k-1} \otimes (\mathrm{diag}(\mathbf{J}_1(\xi_1), \mathbf{J}_2(\xi_2), \ldots, \mathbf{J}_n(\xi_n)) - \varepsilon\mathbf{B}_n \otimes \mathbf{C})\mathbf{U} \atop -\varepsilon\big(\mathbf{D}_{n(k-1)} \otimes \mathbf{C}\big)\mathbf{U}, \tag{6.23}$$

where $\mathbf{U} = (\mathbf{U}_1, \mathbf{U}_2 \ldots \mathbf{U}_{k-1})^T$, and matrix $\mathbf{D}_{n(k-1)}$:

$$\mathbf{D}_{n(k-1)} = \begin{pmatrix} 2\mathbf{C}_* & -\mathbf{C}^* & 0 & & & & \\ -\mathbf{C}_* & 2\mathbf{C}^* & -\mathbf{C}_* & & \mathbf{0} & & \\ 0 & -\mathbf{C}^* & 2\mathbf{C}_* & & & & \\ & & & \ddots & & & \\ & & & & 2\mathbf{C}_* & -\mathbf{C}^* & 0 \\ & \mathbf{0} & & & -\mathbf{C}_* & 2\mathbf{C}^* & -\mathbf{C}_* \\ & & & & 0 & -\mathbf{C}^* & 2\mathbf{C}_* \end{pmatrix}.$$

Applying Eq. (6.23) to Lemma 6.2, we obtain: $\mathbf{A}(t) = \mathbf{I}_{n(k-1)} \otimes \mathrm{diag}(\mathbf{J}_1(\xi_1), \mathbf{J}_2(\xi_2), \ldots, \mathbf{J}_n(\xi_n))$, $\mathbf{B} = \big(\mathbf{I}_{k-1} \otimes \mathbf{B}_n + \mathbf{D}_{n(k-1)}\big) \otimes \mathbf{C}$. As we can see, the norm of the variable matrix remains the same. In addition, the minimum root of matrix $\mathbf{I}_{k-1} \otimes \mathbf{B}_n + \mathbf{D}_{n(k-1)}$ is $\mathbf{l}_{\min} = 4\sin^2 \frac{\mathbf{p}}{2nk}$. Everything together gives stability conditions of the structure, which are formulated in the same form.

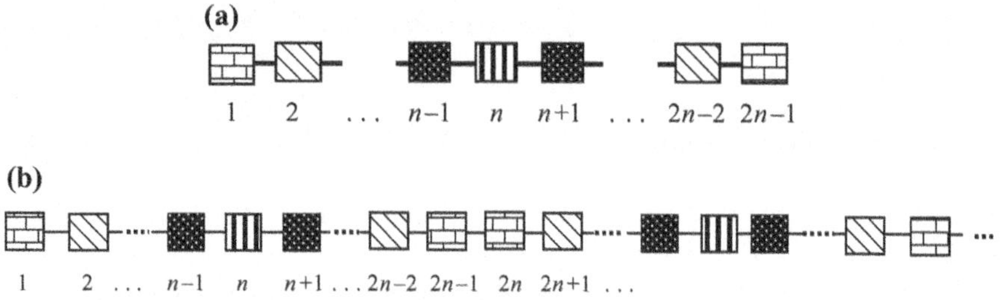

(a)

1   2   ... $n-1$ n $n+1$ ... $2n-2$ $2n-1$

(b)

1   2   ... $n-1$ n $n+1$ ... $2n-2$ $2n-1$ $2n$ $2n+1$ ...

Fig. 6.52 A circuit of the cluster structure of a simple cell (**a**), and a circuit of the structure in the sequential fusion of simple cells, (**b**)

(3) *Stability of cluster structures based on simple cells $O_a(n)$.*

As it is already known, generalized oscillators of the type $O_a(n)$ exist in pairs, forming a simple cell with a certain cluster structure (see Fig. 6.52a).

First of all, we investigate the stability of the structure of the very simple cell, considering its equations as the equations of a chain of elementary oscillators with number $N = 2n - 1$:

$$\dot{\mathbf{X}}_i = \mathbf{F}(\mathbf{X}_i) - \varepsilon\mathbf{C}(-\mathbf{X}_{i-1} + 2\mathbf{X}_i - \mathbf{X}_{i+1}),$$
$$i = \overline{1, N}, \mathbf{X}_0 = \mathbf{X}_1, \mathbf{X}_N = \mathbf{X}_{N+1}. \tag{6.24}$$

Due to symmetry of the cluster structure, we carry out linearization of (6.24) by variables $\mathbf{U}_i = \mathbf{X}_i - \mathbf{X}_{N-i+1}$, $i = \overline{1, n-1}$. A linearized system has the form

$$\dot{\mathbf{U}}_i = \mathbf{J}_i(\xi_i)\mathbf{U}_i - \varepsilon\mathbf{C}(-\mathbf{U}_{i-1} + 2\mathbf{U}_i - \mathbf{U}_{i+1}),$$
$$i = \overline{1, n-1}, \mathbf{U}_0 = \mathbf{U}_1, \mathbf{U}_n = 0. \tag{6.25}$$

System (6.25) written in a form of one equation

$$\dot{\mathbf{U}} = \left(\mathbf{J}^*(\xi) - \varepsilon\mathbf{D}_p \otimes \mathbf{C}\right)\mathbf{U} \tag{6.26}$$

where $\mathbf{J}^*(\xi) = \mathrm{diag}(\mathbf{J}_1(\xi_1), \mathbf{J}_2(\xi_2), \dots, \mathbf{J}_{n-1}(\xi_{n-1}))$ is the truncated Jacobian of a C-oscillator of the type $O_a(n), (\xi_1, \xi_2, \dots, \xi_n) \in A_a(n)$, and matrix $\mathbf{D}_p$ has the form

$$
\mathbf{D}_p =
\begin{Vmatrix}
1 & -1 & 0 & \cdots & \cdots & 0 & 0 \\
-1 & 2 & -1 & \ddots & \cdots & 0 & 0 \\
0 & -1 & 2 & \ddots & \ddots & 0 & 0 \\
\vdots & \ddots & \ddots & \ddots & \ddots & \ddots & \vdots \\
\vdots & \vdots & \ddots & \ddots & 2 & -1 & 0 \\
0 & 0 & 0 & \ddots & -1 & 2 & -1 \\
0 & 0 & 0 & \cdots & 0 & -1 & 2
\end{Vmatrix}.
$$

As one can see, Eq. (6.26) falls completely under Lemma 6.2. The norm of the variable matrix remains the same, and the minimal root of the matrix $\mathbf{D}_p$ has the form $\lambda_{\min} = 4\sin^2 \frac{\pi}{2(2n-1)}$. Thus, stability conditions of a cluster structure of the simple cell has an already known form: $\varepsilon > \mu / \lambda_{\min}$, if $\mathbf{C} = \mathbf{I}$, and $\varepsilon > \varepsilon^*(\mu) / \lambda_{\min}$ in the case of matrix $\mathbf{C}$ of the general form.

We consider structures on the basis of simple cells in parallel and sequential fusion. We assume that the stability conditions of the structure of the simple cell itself are satisfied, and consider the stability of synchronization of the assemblies of these internally structured entities. It is obvious that in parallel fusion (see Fig. 6.53) we obtain the same stability conditions as in the case of a cluster forming oscillator of the type $O_s(n)$, and to adopt them we should just use a change of the form $\lambda_s(n) \rightarrow \lambda_a(n)$.

In the case of a sequential fusion (see Fig. 6.52b), the situation is also simplified due to the symmetry of the internal structure of the simple cell: the matrix of sequential fusion can always be $\mathbf{C}_* = \mathrm{diag}(0, 0, \ldots, 0, 1)$ or $\mathbf{C}^* = \mathrm{diag}(1, 0, \ldots, 0, 0)$ with dimension $2n - 1$; and it does not matter which one of then would it be exactly. Thus, one can use system (6.22) after the change $\mathbf{C}^* \rightarrow \mathbf{C}_*$. In this case, matrix $\mathbf{D}_{(2n-1)(k-1)}$ (change $n \rightarrow 2n - 1$) takes the form $\mathbf{D}_{(2n-1)(k-1)} = \mathbf{D}_{k-1} \otimes \mathbf{C}_*$, where

Fig. 6.53 Cluster structure obtained during a parallel fusion

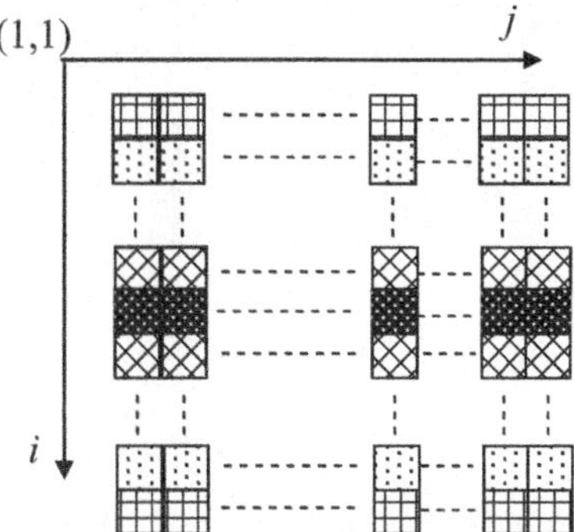

$$\mathbf{D}_{k-1} = \begin{Vmatrix} 2 & -1 & 0 & \cdots & \cdots & 0 & 0 \\ -1 & 2 & -1 & \ddots & \cdots & 0 & 0 \\ 0 & -1 & 2 & \ddots & \ddots & 0 & 0 \\ \vdots & \ddots & \ddots & \ddots & \ddots & \ddots & \vdots \\ \vdots & \vdots & \ddots & \ddots & 2 & -1 & 0 \\ 0 & 0 & 0 & \ddots & -1 & 2 & -1 \\ 0 & 0 & 0 & \cdots & 0 & -1 & 2 \end{Vmatrix}.$$

Taking into account the aforesaid matters, Eq. (6.23) takes the form

$$\dot{\mathbf{U}} = \mathbf{I}_{k-1} \otimes \mathbf{J}(\xi)\mathbf{U} - \varepsilon(\mathbf{D}_{k-1} \otimes \mathbf{C}_* \otimes \mathbf{C})\mathbf{U}, \tag{6.27}$$

where $\mathbf{J}(\xi) = \mathrm{diag}(\mathbf{J}_1(\xi_1), \mathbf{J}_2(\xi_2), \ldots, \mathbf{J}_{2n-1}(\xi_1)) - \varepsilon\mathbf{B}_{2n-1} \otimes \mathbf{C}$ is the Jacobian of the simple cell.

As one can see, Eq. (6.27) falls under Lemma 5.1, and, consequently, the stability problem (6.27) reduces to the corresponding problem for an equation of the form

$$\dot{\mathbf{U}} = \mathbf{J}(\xi)\mathbf{U} - \varepsilon\lambda_{\min}\mathbf{C}_* \otimes \mathbf{C}\mathbf{U}, \tag{6.28}$$

where $\lambda_{\min} = 4\sin^2\frac{\pi}{2k}$ is the minimal root of matrix $\mathbf{D}_{k-1}$.

Equation (6.28) is not the final point of the solution of the problem, since it has the same problem as does Eq. (6.19). Therefore, we extend the transformations (6.28) to Lemma 6.2, performing the same procedures as done for Eq. (6.19).

As a result, we obtain the norm for the variable matrix as the largest of the norms of the partial Jacobians of oscillators of the simple cell, as well as the minimal eigenvalue of the constant matrix–the minimal root $\Lambda_{\min}$ of matrix $\mathbf{B}_{2n-1} + \lambda_{\min}\mathbf{C}_*$:

$$\mathbf{B}_{2n-1} + \lambda_{\min}\mathbf{C}_* = \begin{Vmatrix} 1 & -1 & 0 & \cdots & \cdots & 0 & 0 \\ -1 & 2 & -1 & \ddots & \cdots & 0 & 0 \\ 0 & -1 & 2 & \ddots & \ddots & 0 & 0 \\ \vdots & \ddots & \ddots & \ddots & \ddots & \ddots & \vdots \\ \vdots & \vdots & \ddots & \ddots & 2 & -1 & 0 \\ 0 & 0 & 0 & \ddots & -1 & 2 & -1 \\ 0 & 0 & 0 & \cdots & 0 & -1 & a \end{Vmatrix}, \quad a = 1 + 4\sin^2\frac{\pi}{2k}.$$

In particular, if $k = 2$, then $\Lambda_{\min} = 4\sin^2\frac{\pi}{4(2n-1)}$; if $k = 3$, then $\Lambda_{\min} = 4\sin^2\frac{\pi}{2(4n-1)}$. The cluster structure (see Fig. 6.53) is stable for $\varepsilon > \mu/\lambda_{\min}$, if $\mathbf{C} = \mathbf{I}$, and $\varepsilon > \varepsilon^*(\mu)/\lambda_{\min}$ is the case of matrix $\mathbf{C}$ of a general form.

We emphasize once again that the properties of cluster attractors depend to a large extent on the lattice parameter ε: if one of them is chaotic for a relatively small ε, then with an increase of this parameter, the attractor quickly becomes regular and then disappears. That is, obtained stability conditions for the parameter ε have upper bounds concerned with the exact type of the attractor and its existence. In particular, this entails restrictions on the number of clusters in stable structures. In a typical case, cluster structures with chaotic internal dynamics that exist for "moderate" ε are multistable—stable in one direction in the phase space and unstable in other directions. This leads to the fact that, in a numerical or full-scale experiment, instead of one cluster structure, a kaleidoscope of cluster structures that are interchangeable in time can be observed.

References

1. Josic, K.: Invariant manifolds and synchronization of coupled dynamical systems. Phys. Rev. Lett. **80**, 3053–3056 (1998)
2. Kaneko, K.: Relevance of clustering to biological networks. Phys. D. **75**, 55–73 (1994)
3. Kaneko, K.: Clustering, coding, switching, hierarchical ordering, and control in a network of chaotic elements. Phys. D. **41**, 137–172 (1990)
4. Georgiou, I.T., Bajaj, A.K., Corless, M.: Invariant manifolds and chaotic vibrations in singularly perturbed nonlinear oscillators. Int. J. Eng. Sci. **36**, 431–458 (1998)
5. Belykh, V.N., Belykh, I.V., Hasler, M.: Hierarchy and stability of partially synchronous oscillations of diffusively coupled dynamical systems. Phys. Rev. E **62**, 6332–6345 (2000)
6. Belykh, V.N., Belykh I.V., Mosekilde, E.: Cluster synchronization modes in an ensemble of coupled chaotic oscillators. Phys. Rev. E. **63**, 036216 (2001)
7. Belykh, V.N., Belykh, I.V., Hasler, M., Nevidin, K.: Cluster synchronization in three-dimensional lattices of diffusively coupled oscillators. Int. J. Bifurc. Chaos. **13**, 755–779 (2003)
8. Okuda, K.: Variety and generality of clustering in globally coupled oscillators. Phys. D. **63**, 424–436 (1993)
9. Xie, F., Hu, G.: Phys. Rev. E **55**, 79 (1997)
10. Hasler, M., Maistrenko, Yu., Popovich, O.: Phys. Rev. E **58**, 6843 (1998)
11. Zanette, D.H., Mikailov, A.S.: Phys. Rev. E **57**, 276 (1998)
12. Verichev, N.N.: C-oscillators and new outlook on cluster dynamics. J. Phys. Conf. Ser. **23**, 23–46 (2005)
13. Verichev, N.N., Verichev, S.N., Wiercigroch, M.: Physical interpretation and theory of existence of cluster structures in lattices of dynamical systems. Chaos, Solitons Fractals **34**(4), 1082–1104 (2007)
14. Verichev, N.N.: Physics, existence and fusion of cluster structures of coupled dynamical systems. Nonlinear World **7**(1), 28–45 (2009)
15. Chua, L.O., Komuro, M., Matsumoto, T.: The double scroll family. IEEE Trans. Circ. Syst. **CAS-33** (11), 1073–1118 (1986)
16. Special Issue on Chua's Circuit. J. Circ. Syst. Comput. **3**(2) (1993)
17. Potapov, A.A.: Fractals in Radiophysics and Electromagnetic Detections. Logos, Moscow (2002)
18. Mandelbrot, B.B.: The Fractal Geometry of Nature. Freeman, New York (1982)
19. Potapov, A.A., Gilmutdinov A.H., Ushakov P.A.: Fractional-Order Radio-Elements and Radio-Systems—Radiotechnika 200 p. (2009) (in Russian)

20. Vonsovski, S.V.: Modern Scientific Picture of the World. Publishing House of the RHD (2006) (in Russian)
21. Agol, V.I., Bogdanovm A.A., Gvozdev V.A., etc. Molecular biology. In: Spirin, A.S. (ed.) Structure and Biosynthesis of Nucleic Acids. Higher School, Moscow (1990)
22. Pecora, L.M., Carroll, T.L.: Synchronization in chaotic systems. Phys. Rev. Lett. **64**, 821 (1990)
23. Osipov, G.V., Sushchik, M.M.: Synchronized clusters and multistability in arrays of oscillators with different natural frequencies. Phys. Rev. E. **58**, 7198 (1998)
24. Kanakov, O.I., Osipov, G.V., Chan, C.-K., Kurths, J.: Cluster synchronization and spatio-temporal dynamics in networks of oscillatory and excitable Luo-Rudy cells. Chaos, Solitons Fractals. **17**, 015111 (2007)
25. Pecora, L.M., Carroll, T.L.: Phys. Rev. Lett. **80**, 2109–2112 (1998)
26. Verichev, N.N., Verichev, S.N., Erofeev, V.I.: Cluster dynamics of a homogeneous chain of dissipatively coupled rotators. Appl. Math. Mech. **72**(6), 882–897 (2008)
27. Rabinovich, M.I., Trubetzkov, D.I.: Introduction in theory of oscillations and waves. Nauka, Moscow (1984). (in Russian)
28. Afraimovich, V.S., Nekorkin, V.I., Osipov, G.V., Shalfeev, V.D.: Stability, structures and chaos in nonlinear synchronization networks. In: Gaponov-Grekhov, A.V., Rabinovich, M.I. (eds.). IPF Academy of Sciences of the USSR, Gorky (1989) (in Russian)
29. Mosekilde, E., Maistrenko, Y., Postnov, D.: Chaotic Synchronization: Applications to Living Systems. World Scientific, Singapore (2002)
30. Wang, W., Kiss, I.Z., Hudson, J.L.: Chaos **10**, 248 (2000)
31. Roy, R., Thornburg, K.S.: Phys. Rev. Lett. **72**, 2009 (2000)
32. Cuomo, K.M., Oppenheim, A.V.: Phys. Rev. Lett. **71**, 65 (1993)
33. Verichev, N.N., Verichev, S.N., Erofeev, V.I.: C-oscillators in a homogeneous ring of diffusive-coupled dynamical systems: existence, stability, and fusion of cluster structures. Nonlinear World. **6** (56), 398–423 (2008) (in Russian)
34. Gantmacher, F.R.: The Theory of Matrices. Chelsea Pub. Co. (1960)
35. Bylov, F., Vinograd, R.E., Grobman, D.M.: The Theory of Lyapunov exponents and Its Applications to Stability Problems. Nauka, Moscow (1966) (in Russian)
36. Myshkis, A.D.: Mathematics. Special Courses. Nauka, Moscow (1971) (in Russian)

Appendix A
Algorithms of Transformation of Systems of Coupled Rotators to the Standard Form

A.1 Systems of Coupled Rotators Without Additional Loads

Consider a case when each of the rotators is not loaded by any additional aperiodic or oscillatory loads. In this case, system (A1.1) has the form

$$I\ddot{\varphi}_k + \delta_k(1 + F_{1k}(\varphi_k))\dot{\varphi}_k + F_{2k}(\varphi_k)$$

$$= \gamma_k + \sum_{j=1}^{n} f_{kj}(\varphi_j) + \sum_{j=1}^{n} \beta_{kj}\dot{\varphi}_j + f_k(\psi) + \mu F_k^*(\varphi),$$

$$\dot{\psi} = \omega_0, \tag{A1.1}$$

where $\langle f_{kj}(\varphi_j)\rangle_{\varphi_j} = 0, \quad \langle f_k(\psi)\rangle_{\psi} = 0, \quad \mu = I^{-1}$.

In other words, we consider a system of rotators coupled by angular speeds and angular accelerations that results in couplings by periodic functions and "other couplings" expressed by terms $\mu F_k^*(\varphi)$.

We need to determine a form of transformation $(\varphi, \dot{\varphi}) \to (\varphi, \xi)$, which reduces system (A1.1) to a system in a standard form with fast spinning phases

$$\dot{\xi}_k = \mu\, \Xi_k(\varphi, \psi, \xi),$$

$$\dot{\varphi}_k = \omega_k + \mu\Phi_k(\varphi, \psi, \xi_k),$$

$$\dot{\psi} = \omega_0, \tag{A1.2}$$

where ω_k, $\Xi_k(\varphi, \psi, \xi)$, $\Phi_k(\varphi, \psi, \xi_k)$ are the unknown parameters and functions to be determined.

Functions $\Xi_k(\varphi, \psi, \xi)$ and $\Phi_k(\varphi, \psi, \xi_k)$ are bounded by all phase variables (there are no secular terms for φ), which correspond to their periodicity by these variables.

© Springer Nature Switzerland AG 2020
N. Verichev et al., *Chaos, Synchronization and Structures in Dynamics of Systems with Cylindrical Phase Space*, Understanding Complex Systems,
https://doi.org/10.1007/978-3-030-36103-7

Substitution of the second equation from system (A1.2) into (A1.1) leads to an equation that determines the sought parameters and functions of the transformation:

$$\sum_{j=1}^{n} \frac{\partial \Phi_k}{\partial \varphi_j}\left(\omega_j + \mu \Phi_j\right) + \frac{\partial \Phi_k}{\partial \psi}\omega_0 + \dot{\xi}_k + \delta_k(1 + F_{1k})(\omega_k + \mu \Phi_k) + F_{2k}$$

$$= \gamma_k + \sum_{j=1}^{n} f_{kj} + \sum_{j=1}^{n} \beta_{kj}\left(\omega_j + \mu \Phi_j\right) + f_k(\psi) + \mu F_k^*(\varphi). \tag{A1.3}$$

Using the condition of boundedness, we obtain the equations for normal frequencies of rotators:

$$\delta_k \omega_k - \sum_{j=1}^{n} \beta_{kj}\omega_j = \gamma_k, \quad k = \overline{1, n}. \tag{A1.4}$$

These equations are obtained after the elimination of all constants from (A1.3). Taking into account the form of the first equation in system (A1.2) $\left(\dot{\xi} \sim \mu\right)$, we obtain equations for determining functions Φ_k (elimination of terms not proportional to μ):

$$\sum_{j=1}^{n} \frac{\partial \Phi_k}{\partial \varphi_j}\omega_j + \frac{\partial \Phi_k}{\partial \psi}\omega_0 + \delta_k \omega_k F_{1k}(\varphi_k) + F_{2k}(\varphi_k)$$

$$= \sum_{j=1}^{n} f_{kj}\left(\varphi_j\right) + f_k(\psi). \tag{A1.5}$$

Linear system (A1.4) is easy to solve. Equation (A1.5) is also simple. The sought solution has the form

$$\Phi_k = \sum_{j=1}^{n} \frac{1}{\omega_j} \int f_{kj}\left(\varphi_j\right)d\varphi_j + \frac{1}{\omega_0} \int f_k(\psi)d\psi - \delta_k \int F_{1k}(\varphi_k)d\varphi_k$$

$$- \frac{1}{\omega_k} \int F_{2k}(\varphi_k)d\varphi_k + \xi_k.$$

These functions are bounded by phase variables (this is a result of the elimination of constants). For functions Ξ_k, we obtain expressions

$$\Xi_k = \sum_{j=1}^{n}\left(\beta_{kj} - \frac{\partial \Phi_k}{\partial \varphi_j}\right)\Phi_j - \delta_k \Phi_k(1 + F_{1k}) + F_k^*.$$

As it was already said, for the interacting quasi-linear systems, the "strong" resonances are the harmonic resonances corresponding to an integer ratio of the frequencies. By saying "strong", we mean that resonance takes place in an averaged system in the first approximation. According to this terminology, subharmonic resonances are "weak". Generally, the most significant is the first (main) resonance. We consider self-oscillating systems and, therefore, the main resonance corresponds to the "simple" mutual synchronization of oscillations and all other resonances correspond to a multiple ("divisible") synchronization. The same can be said with respect to the master-slave synchronization of a self-oscillatory system by an external force. To study harmonic resonances in system (A1.2), we introduce phase and frequency mistunings of the form $q_k \varphi_k - p_k \psi = \eta_k$, $q_k \omega_k - p_k \omega_0 = \mu \Delta_k$, where $p_k/q_k = n_k$ are integers. The value of $n_k = 1$ corresponds to the simple synchronization. As a result, we obtain a system with one fast spinning phase ψ of the form

$$
\begin{aligned}
\dot{\xi}_k &= \mu \Xi_k(\eta, \psi, \xi), \\
\dot{\eta}_k &= \mu(\Delta_k + q_k \Phi_k(\eta, \psi, \xi_k)), \\
\dot{\psi} &= \omega_0.
\end{aligned}
\tag{A1.6}
$$

We apply the method of the averaging to obtain:

$$
\Xi_k^*(\eta, \mathbf{x}) = \frac{1}{2\pi} \int_0^{2\pi} \Xi_k(\eta, \mathbf{x}, \psi) \partial \psi,
$$

$$
\frac{1}{2\pi} \int_0^{2\pi} (\Delta_k + q_k \Phi_k) \partial \psi = \Delta_k + q_k \xi_k
$$

and an averaged system of the form

$$
\begin{aligned}
\dot{\xi}_k &= \Xi_k^*(\eta, \xi), \\
\dot{\eta}_k &= \Delta_k + q_k \xi_k.
\end{aligned}
\tag{A1.7}
$$

In system (A1.7), we have kept old names for the averaged variables and transformed time $\mu \tau = \tau_n$. By studying system (A1.7), one can particularly obtain rotation characteristics of rotators in the resonance zones: $\langle \dot{\varphi}_k \rangle_\tau = \Omega_k = n_k \omega_0 + \frac{\mu}{q_k} \langle \dot{\eta}_k^* \rangle_\tau$.

During the interpretation of dynamical properties of averaged system (A1.7) onto the original system (A1.1), we take into account that if G is a limit set of trajectories of the averaged system, then $G \times S^1$ is a limit set of trajectories of the original system with some limitations described above. Also, one needs to remember the relation between the trajectories of the averaged system and the trajectories of the point mapping (Poincaré mapping) for the original system.

Remark 1 The choice of a "fast" phase for the averaging in system (A1.2) is normally done according to considerations of convenience. If a system is non-autonomous, then such phase is a phase of external force ψ. If systems (A1.1) and (A1.2), respectively, are autonomous, then any phase can be chosen. If such phase is $\varphi_1 = \varphi$, then phase and frequency mistunings of the form $q_k\varphi_k - p_k\varphi = \eta_k$, $q_k\omega_k - p_k\omega_1 = \mu\Delta_k$, where p, q are integers, are to be introduced to study the synchronization.

Remark 2 If system (A1.1) is non-autonomous, then during the transformations it is convenient to choose a form of equations for the phases of rotators in the form $\dot{\varphi}_k = n_k\omega_0 + \mu\Phi_k(\boldsymbol{\varphi}, \boldsymbol{\psi}, \xi_k)$. If the system is autonomous, then $\dot{\varphi}_k = n_k\omega + \mu\Phi_k(\boldsymbol{\varphi}, \xi_k)$, where ω is a normal frequency of the "averaging" rotator, and n_k are the integers. In this case, frequency mistunings pass into the equation for variable $\boldsymbol{\xi}$. Note that the choice of the frequency does not affect the result, but reduces the procedure of transformations.

A.2 Systems of Coupled Rotators with Aperiodic Loads

First we consider one rotator with an aperiodic load, the equations of which have an arbitrary dimension. Suppose that matrix $\mathbf{A}$ has real roots and, without loss of generality, we consider that it is diagonal. We also assume that the load has a small dissipation, i.e. $\mathbf{A} = \mu\mathbf{B}$, where $\mathbf{B}$ is a diagonal nondegenerate matrix. Under the above assumptions, we consider a dynamical system of the form

$$I\ddot{\varphi} + \delta(1 + F_1(\varphi))\dot{\varphi} + F_2(\varphi) = \gamma + \mathbf{a}^T\mathbf{z} + f(\psi) + \mu F^*(\varphi, \mathbf{z}),$$
$$\dot{\mathbf{z}} = \mu(\mathbf{B}\mathbf{z} + \mathbf{b}\dot{\varphi} + \mathbf{c}f_1(\varphi) + \mathbf{d}f_2(\psi)), \tag{A2.1}$$

where $\mathbf{a}$, $\mathbf{b}$, $\mathbf{c}$, and $\mathbf{d}$ are the constant vectors.

It is easy to see that a "generating" solution for the second equation (equation for the load; solution for $\mu = 0$) is solution $\mathbf{z} = \text{const}$. This means that if one would transform the equation for the rotator to the standard form directly from (A2.1), then variable $\mathbf{z}$ will be included in to the "generating" frequency of the rotator. This circumstance will complicate the entire procedure of the averaging. The ideal case is a case when a "generating" frequency does not depend on slow variables (i.e. it is a parameter). Due to this reason, let us perform the following transformation of the equation for the load:

$$\mathbf{z} = \boldsymbol{\beta} + \mu\boldsymbol{\Psi}(\varphi, \psi) + \mu\mathbf{y}, \tag{A2.2}$$

where constant vector $\boldsymbol{\beta}$ and vector-function $\boldsymbol{\Psi}(\varphi, \psi)$ are to be determined. Substituting Eq. (A2.2) into the second equation of (A2.1) and taking into account that the equation for phase φ has the form $\dot{\varphi} = \omega_0 + \mu\Phi$, we obtain:

$$\frac{\partial\boldsymbol{\Psi}}{\partial\varphi}(\omega_0 + \mu\Phi) + \frac{\partial\boldsymbol{\Psi}}{\partial\psi}\omega_0 + \dot{\mathbf{y}}$$
$$= \mu\mathbf{B}(\boldsymbol{\Psi} + \mathbf{y}) + \mathbf{B}\boldsymbol{\beta} + \mathbf{b}\omega_0 + \mu\mathbf{b}\Phi + \mathbf{c}f_1(\varphi) + \mathbf{d}f_2(\psi). \tag{A2.3}$$

Equations that determine $\boldsymbol{\beta}$ and $\mathbf{Y}(\varphi, \psi)$, are obtained using the following simple considerations: first, the equation for variable $\mathbf{y}$ has to have a standard form; second, function $\mathbf{Y}(\varphi, \psi)$, has to be bounded. For these conditions, the sought equations have the form

$$\frac{\partial \boldsymbol{\Psi}}{\partial \varphi} \omega_0 + \frac{\partial \boldsymbol{\Psi}}{\partial \psi} \omega_0 = \mathbf{c} f_1(\varphi) + \mathbf{d} f_2(\psi),$$

$$\mathbf{B}\boldsymbol{\beta} + \mathbf{b}\omega_0 = 0. \tag{A2.4}$$

Solving Eq. (A2.4) is not difficult:

$$\boldsymbol{\Psi} = \frac{\mathbf{c}}{\omega_0} \int f_1(\varphi) d\varphi + \frac{\mathbf{d}}{\omega_0} \int f_2(\psi) d\psi,$$

$$\boldsymbol{\beta} = -\mathbf{B}^{-1} \mathbf{b}\omega_0.$$

For the found solutions, Eq. (A2.1) take the form

$$I\ddot{\varphi} + \delta(1 + F_1(\varphi))\dot{\varphi} + F_2(\varphi) = \gamma + \mathbf{a}^T \boldsymbol{\beta} + f(\psi) + \mu \mathbf{a}^T (\boldsymbol{\Psi} + \mathbf{y}) + \mu F^*(\varphi, \mathbf{z}),$$

$$\dot{\mathbf{y}} = \mu \left(\mathbf{B}\mathbf{y} + \mathbf{B}\boldsymbol{\Psi} - \frac{\partial \boldsymbol{\Psi}}{\partial \varphi} \boldsymbol{\Phi} + \mathbf{b}\boldsymbol{\Phi} \right). \tag{A2.5}$$

And now let us transform the equation for the rotator. Substituting into the first equation of (A2.5) the equation for the phase $\dot{\varphi} = \omega_0 + \mu \Phi(\varphi, \psi, \xi)$, we obtain

$$\frac{\partial \Phi}{\partial \varphi}(\omega_0 + \mu \Phi) + \frac{\partial \Phi}{\partial \psi} \omega_0 + \frac{\partial \Phi}{\partial \xi} \dot{\xi} + \delta(1 + F_1(\varphi))(\omega_0 + \mu \Phi) + F_2(\varphi)$$

$$= \gamma + \mathbf{a}^T \boldsymbol{\beta} + f(\psi) + \mu \mathbf{a}^T (\boldsymbol{\Psi} + \mathbf{y}) + \mu F^*(\varphi, \mathbf{z}). \tag{A2.6}$$

The equation for function $\Phi(\varphi, \psi, \xi)$ is obtained using the same considerations as those for (I.1):

$$\frac{\partial \Phi}{\partial \varphi} \omega_0 + \frac{\partial \Phi}{\partial \psi} \omega_0 + \delta \omega_0 F_1(\varphi) + F_2(\varphi) = f(\psi).$$

Its solutions are: $\Phi = -\delta \int F_1(\varphi) d\varphi - \frac{1}{\omega_0} \int F_2(\varphi) d\varphi + \frac{1}{\omega_0} \int f(\psi) d\psi + \xi$.

Relation $\gamma + \mathbf{a}^T \boldsymbol{\beta} - \delta \omega_0 = \mu \Delta$ determines the resonance zone of the parameters. Eventually, we obtain system of equations in a standard form, which is equivalent to system (A2.1):

$$\dot{\xi} = \mu \Xi,$$

$$\dot{\mathbf{y}} = \mu \mathbf{Y},$$

$$\dot{\eta} = \mu \Phi,$$

$$\dot{\varphi} = \omega_0 + \mu \Phi, \tag{A2.7}$$

where $\Xi = \Delta - \delta(1 + F_1(\varphi))\Phi - \frac{\partial \Phi}{\partial \varphi}\Phi + \mathbf{a}^T(\mathbf{\Psi} + \mathbf{y}) + F^*(\varphi, \mathbf{z})$, $\mathbf{Y} = \mathbf{By} + \mathbf{B}\mathbf{\Psi} - \frac{\partial \mathbf{\Psi}}{\partial \varphi}\Phi + \mathbf{b}\Phi$, $\eta = \varphi - \psi$ is the phase mistuning.

Remark 3 We have considered the case of a non-autonomous system. Here, the generating frequency has been determined and it was a frequency of external force ω_0. In the autonomous case, the generating frequency of the rotator is a priori unknown. Formally, it is found using the resonance relation. Namely, assuming $\Delta = 0$, we obtain the generating frequency of the rotator: $\omega_0 = \frac{\gamma}{(\mathbf{a}^T\mathbf{B}^{-1}\mathbf{b}+\delta)}$.

Consider a system of an arbitrary number of coupled rotators with aperiodic loads. Suppose that the equations for all aperiodic loads are transformed in the aforementioned way and the system of coupled rotators has the form

$$I\ddot{\varphi}_k + \delta(1 + F_{1k}(\varphi_k))\dot{\varphi}_k + F_{2k}(\varphi_k) = \gamma_k + \mathbf{a}_k^T\boldsymbol{\beta}_\kappa + f_k(\psi) + \sum_{j=1}^{n} f_{kj}(\varphi_j)$$

$$+ \sum_{j=1}^{n}\beta_{kj}\dot{\varphi}_j + \mu\mathbf{a}_k^T(\mathbf{\Psi}_k + \mathbf{y}_k) + \mu F_k^*(\varphi, \mathbf{z}),$$

$$\dot{\mathbf{y}}_k = \mu\mathbf{Y}_k,$$

$$\dot{\psi} = \omega_0. \tag{A2.8}$$

Transformation of the equations for the rotators in system (A2.8) is the same as that for system (A1.1). Moreover, since the impact of the loads on the rotators is "weak" in system (A2.8), then functions Φ_k will not depend on vectors y_k. They will have the same expressions as in Section (A1.1).

Performing transformations and determining the necessary functions, we obtain the system in a standard form equivalent to system (A2.8) that has the form

$$\dot{\boldsymbol{\xi}}_k = \mu\Xi_k(\boldsymbol{\varphi}, \psi, \boldsymbol{\xi}, \mathbf{y}_k),$$

$$\dot{\mathbf{y}}_k = \mu\mathbf{Y}_k(\boldsymbol{\varphi}, \psi, \boldsymbol{\xi}, \mathbf{y}_k),$$

$$\dot{\varphi}_k = \omega_k + \mu\Phi_k(\boldsymbol{\varphi}, \psi, \boldsymbol{\xi}_k),$$

$$\dot{\psi} = \omega_0. \tag{A2.9}$$

The equations and, consequently, expressions for functions Φ_k are the same as those in (A1.1):

$$\Phi_k = \sum_{j=1}^{n}\frac{1}{\omega_j}\int f_{kj}(\varphi_j)d\varphi_j + \frac{1}{\omega_0}\int f_k(\psi)d\psi - \delta_k\int F_{1k}(\varphi_k)d\varphi_k$$

$$- \frac{1}{\omega_k}\int F_{2k}(\varphi_k)d\varphi_k + \xi_k.$$

Expressions for Ξ_k and Y_k:

$$\Xi_k = \sum_{j=1}^{n} \left(\beta_{kj} - \frac{\partial \Phi_k}{\partial \varphi_j} \right) \Phi_j - \delta_k \Phi_k (1 + F_{1k}) + F_k^* + \mathbf{a}_k^T (\mathbf{\Psi}_k + \mathbf{y}_k),$$

$$\mathbf{Y}_k = \mathbf{B}_k \mathbf{y}_k + \mathbf{B}_k \mathbf{\Psi}_k - \frac{\partial \mathbf{\Psi}_k}{\partial \varphi_k} \Phi_k + \mathbf{b}_k \Phi_k$$

Consequently, expressions for the generating frequencies of the rotators have the form

$$\left(\delta_k + \mathbf{a}_k^T \mathbf{B}_k^{-1} \mathbf{b}_k \right) \omega_k - \sum_{j=1}^{n} \beta_{kj} \omega_j = \gamma_k,$$

$$k = \overline{1, n}.$$

Further, taking into account Remarks 1 and 2, system (A2.9) is to be reduced to the system in a standard form with one fast spinning phase and the procedure of the averaging is to be carried out.

A.3　Systems of Coupled Rotators with Oscillatory Loads

First, let us consider a case of one rotator interacting with a multifrequency oscillatory load.

As it was said above, a quasi-linear rotator effectively interacts with an oscillatory load only in the resonance case, when the oscillator has a high quality (small damping). We will suppose that all dissipative terms of oscillators are described by function $\mathbf{Z}(\mathbf{z}, \varphi, \dot{\varphi}, \psi)$ of system (A1.1). We will also assume that the ratio of all frequencies of oscillators to the least frequency (or to the frequency of external impact) is close to an integer, i.e. $\omega_j^2 = n_j^2 \omega_0^2 + \mu \nu_j$, where n_j are integers. Suppose that all terms of the equation concerned with frequency mistunings $(\mu \nu_j)$, are also described by function $\mathbf{Z}(\mathbf{z}, \varphi, \dot{\varphi}, \psi)$ Also, we assume that the vector equation for an oscillatory system is transformed to the equilibrium in the origin. The named conditions are easy to fulfil using elementary transformations of equations of an oscillatory system and we assume that these transformations are performed. The systems considered throughout this work can serve as examples.

Taking these into account, consider a dynamical system of the form

$$I\ddot{\varphi} + \delta(1 + F_1(\varphi))\dot{\varphi} + F_2(\varphi) = \gamma + \mathbf{a}^T \mathbf{z} + f(\psi) + \mu F^*(\varphi, \mathbf{z}),$$

$$\dot{z} = \mathbf{A}\mathbf{z} + \mu \mathbf{Z}(\mathbf{z}, \varphi),$$

$$\dot{\psi} = \omega_0, \tag{A3.1}$$

where $\mathbf{a}$ is a constant vector and matrix $\mathbf{A}$ has imaginary eigenvalues and Jordan form with cells of the form $\mathbf{A}_j = \begin{pmatrix} 0 & 1 \\ -n_j^2\omega_0^2 & 0 \end{pmatrix}$, $j = \overline{1, m}$, $\dim(\mathbf{z}, \mathbf{Z}) = 2m$.

It is easy to see that the generating vector equation for the oscillatory system (equation for $\mu = 0$) represents a system of independent harmonic oscillators. Let us rename the variables: $z_{2j-1} = x_j$, $Z_{2j-1} = X_j$, $z_{2j} = y_j$, $Z_{2j} = Y_j$.

According to Remark 2, let us determine the form of the equation for the phase of the rotator using the form $\dot{\varphi} = \omega_0 + \mu\Phi$ and transform all quasi-linear equations of oscillators using a modified van der Pol change in the same manner. In each system of the form

$$\dot{x} = y + \mu X, \quad \dot{y} = -n^2\omega_0^2 x + \mu Y$$

we perform a change of the variables of the form

$$x = \theta \sin n\varphi + \eta \cos n\varphi, \quad y = (\theta \cos n\varphi - \eta \sin n\varphi)n\omega_0,$$

As a result, we obtain the system in a standard form

$$\dot{\theta} = \mu\Theta, \quad \dot{\eta} = \mu T,$$

where

$$\Theta = n\eta\,\Phi + X\,\sin n\varphi + \frac{1}{n\omega_0} Y\,\cos n\varphi,$$

$$T = -n\theta\,\Phi + X\,\cos n\varphi - \frac{1}{n\omega_0} Y\,\sin n\varphi.$$

In the new variables, the equations for the oscillatory system have a standard form with respect to $\theta = (\theta_1, \ldots, \theta_m)^T$, and $\eta = (\eta_1, \ldots, \eta_m)^T$, while system (A3.1) takes the form

$$I\ddot{\varphi} + \delta(1 + F_1(\varphi))\dot{\varphi} + F_2(\varphi) = \gamma + b^T x + c^T y + f(\psi) + \mu F^*(\varphi, z),$$
$$\dot{\boldsymbol{\theta}} = \mu\boldsymbol{\Theta},$$
$$\dot{\boldsymbol{\eta}} = \mu\mathbf{T},$$
$$\dot{\psi} = \omega_0, \tag{A3.2}$$

where $\quad \mathbf{b}^T\mathbf{x} \quad = \quad \sum_{i=1}^{m} b_i(\theta_i \sin n_i\varphi + \eta_i \cos n_i\varphi), \quad \mathbf{c}^T\mathbf{y} \quad = $

$\sum_{i=1}^{m} c_i(\theta_i \cos n_i\varphi - \eta_i \sin n_i\varphi)\,n_i\omega_0;\; b_i, c_i$ are the odd and even coordinates of vector $\mathbf{a}$.

Now, let us transform the equation for the rotator. Substituting the expression for phase $\dot{\varphi} = \omega_0 + \mu \Phi(\varphi, \psi, \xi, \boldsymbol{\theta}, \boldsymbol{\eta})$ into the first equation of system (A3.2), we obtain:

$$\frac{\partial \Phi}{\partial \varphi}(\omega_0 + \mu \Phi) + \frac{\partial \Phi}{\partial \psi}\omega_0 + \sum_{j=1}^{m} \frac{\partial \Phi}{\partial \theta_j}\mu \Theta_j + \sum_{j=1}^{m} \frac{\partial \Phi}{\partial \eta_j}\mu T_j + \dot{\xi}$$

$$+ \delta(1 + F_1)(\omega_0 + \mu \Phi) + F_2 = \gamma + \mathbf{b}^T \mathbf{x} + \mathbf{c}^T \mathbf{y} + f(\psi) + \mu F^*(\varphi, \mathbf{z}).$$

$$\text{(A3.3)}$$

Keeping in mind requirements for function Φ, using (A3.3) we obtain the equation for determining this function

$$\frac{\partial \Phi}{\partial \varphi}\omega_0 + \frac{\partial \Phi}{\partial \psi}\omega_0 + \delta \omega_0 F_1 + F_2 = \mathbf{b}^T \mathbf{x} + \mathbf{c}^T \mathbf{y} + f(\psi) \qquad \text{(A3.4)}$$

And the resonance zone of parameters $\gamma - \delta \omega_0 = \mu \Delta$.
The solution of Eq. (A3.4) has the form:

$$\Phi = -\delta \int F_1(\varphi)d\varphi - \frac{1}{\omega_0} \int F_2(\varphi)d\varphi + \frac{1}{\omega_0} \int f(\psi)d\varphi$$

$$+ \frac{1}{\omega_0} \int \mathbf{b}^T \mathbf{x}(\varphi)d\varphi + \frac{1}{\omega_0} \int \mathbf{c}^T \mathbf{y}(\varphi)d\varphi + \xi,$$

where $\frac{1}{\omega_0}\int \mathbf{b}^T \mathbf{x}(\varphi)d\varphi = \frac{1}{\omega_0}\sum_{i=1}^{m}\frac{b_i}{n_i}(-\theta_i \cos n_i \varphi + \eta_i \sin n_i \varphi)\frac{1}{\omega_0}\int \mathbf{c}^T \mathbf{y}(\varphi)d\varphi = \sum_{i=1}^{m} c_i(\theta_i \sin n_i \varphi + \eta_i \cos n_i \varphi)$.
The expression for function Ξ:

$$\Xi = \Delta - \frac{\partial \Phi}{\partial \varphi}\Phi - \delta(1 + F_1)\Phi - \sum_{j=1}^{m}\frac{\partial \Phi}{\partial \theta_j}\Theta_j - \sum_{j=1}^{m}\frac{\partial \Phi}{\partial \eta_j}T_j + F^*(\varphi, \mathbf{z}).$$

Eventually, we obtain the system in a standard form, which is equivalent to system (A3.1):

$$\dot{\xi} = \mu \Xi,$$
$$\dot{\boldsymbol{\theta}} = \mu \boldsymbol{\Theta},$$
$$\dot{\boldsymbol{\eta}} = \mu \mathbf{T},$$
$$\dot{\chi} = \mu \Phi,$$
$$\dot{\varphi} = \omega_0 + \mu \Phi, \qquad \text{(A3.5)}$$

where $\chi = \varphi - \psi$ is the phase mistuning.

Consider now the system of coupled rotators, where each of them is loaded by an oscillatory load. Equations for loads can have different dimensions and frequency spectra. We will suppose that all of the aforementioned conditions for oscillatory systems are satisfied and their equations are transformed to the standard form.

Consider a dynamical system of the form

$$I\ddot{\varphi}_k + \delta_k(1 + F_{1k}(\varphi_k))\dot{\varphi}_k + F_{2k}(\varphi_k) = \gamma_k + f_k(\psi)$$

$$+ \sum_{j=1}^{n} f_{kj}(\varphi_j) + \sum_{j=1}^{n} \beta_{kj}\dot{\varphi}_j + \mathbf{b}_k^T\mathbf{x}_k + \mathbf{c}_k^T\mathbf{y}_k + \mu F_k^*(\varphi, \mathbf{z}),$$

$$\dot{\boldsymbol{\theta}}_k = \mu\boldsymbol{\Theta}_k,$$

$$\dot{\boldsymbol{\eta}}_k = \mu\mathbf{T}_k,$$

$$\dot{\psi} = \omega_0, \tag{A3.6}$$

where $\mathbf{b}_k^T\mathbf{x}_k = \sum_{i=1}^{m_k} b_{ki}(\theta_{ki}\sin n_{ki}\varphi + \eta_{ki}\cos n_{ki}\varphi)$, $\mathbf{c}_k^T\mathbf{y}_k = \sum_{i=1}^{m_k} c_{ki}(\theta_{ki}\cos n_{ki}\varphi - \eta_{ki}\sin n_{ki}\varphi)n_{ki}\omega_0$, φ is one of the phases of the rotators.

Having defined a form of the equations for the phases $\dot{\varphi}_k = \omega_0 + \mu\Phi_k(\varphi, \psi, \boldsymbol{\theta}_k, \boldsymbol{\eta}_k, \xi_k)$ and substituting these expressions into the equations for rotators, we obtain equations for functions $\Phi_k(\varphi, \psi, \boldsymbol{\theta}_k, \boldsymbol{\eta}_k, \xi_k)$, expressions for functions Ξ_k, and resonance zones of parameters

$$\sum_{j=1}^{n} \frac{\partial\Phi_k}{\partial\varphi_j}\omega_0 + \frac{\partial\Phi_k}{\partial\psi}\omega_0 + \delta_k\omega_0 F_{1k}(\varphi_k) + F_{2k}(\varphi_k)$$

$$= \sum_{j=1}^{n} f_{kj}(\varphi_j) + f_k(\psi) + \mathbf{b}_k^T\mathbf{x}_k + \mathbf{c}_k^T\mathbf{y}_k,$$

$$\gamma_k - \left(\delta_k + \mathbf{a}_k^T\mathbf{B}_k^{-1}\mathbf{b}_k - \sum_{j=1}^{n}\beta_{kj}\right)\omega_0 = \mu\Delta_k, \quad k = \overline{1, n}. \tag{A3.7}$$

The sought solution of Eq. (A3.7) is not difficult to find:

$$\Phi_k = \frac{1}{\omega_0}\sum_{j=1}^{n}\int f_{kj}(\varphi_j)d\varphi_j + \frac{1}{\omega_0}\int f_k(\psi)d\psi - \delta_k\int F_{1k}(\varphi_k)d\varphi_k$$

$$- \frac{1}{\omega_0}\int F_{2k}(\varphi_k)d\varphi_k + \frac{1}{\omega_0}\int \mathbf{b}_k^T\mathbf{x}_k(\varphi)\,d\varphi + \frac{1}{\omega_0}\int \mathbf{c}_k^T\mathbf{y}_k(\varphi)\,d\varphi + \xi_k.$$

The expressions for functions Ξ_k:

$$\Xi_k = \Delta_k + \sum_{j=1}^{n}\left(\beta_{kj} - \frac{\partial\Phi_k}{\partial\varphi_j}\right)\Phi_j - \delta_k\Phi_k(1 + F_{1k})$$

$$-\sum_{j=1}^{m} \frac{\partial \Phi_k}{\partial \theta_{kj}} \Theta_{kj} - \sum_{j=1}^{m} \frac{\partial \Phi_k}{\partial \eta_{kj}} T_{kj} + F_k^*.$$

Eventually, we obtain a system in a standard form with a fast spinning phase, which is equivalent to Eq. (A3.6):

$$\dot{\xi}_k = \mu \, \Xi_k,$$
$$\dot{\boldsymbol{\theta}}_k = \mu \boldsymbol{\Theta}_k,$$
$$\dot{\boldsymbol{\eta}}_k = \mu \mathbf{T}_k,$$
$$\dot{\chi}_k = \mu (\Phi - \Phi_k),$$
$$\dot{\varphi} = \omega_0 + \mu \Phi,$$

where $\chi_k = \varphi - \varphi_k$ are the phase mistunings, and φ is one of the phases of rotators.

Appendix B
Calculation of Eigenvalues of Matrices

Consider matrix $\mathbf{A}_p$ and determinant $\boldsymbol{\Delta}_p$ of the form

$$
\mathbf{A}_p =
\begin{pmatrix}
2v+\alpha & -1 & 0 & 0 & & & & & \\
-1 & 2v & -1 & 0 & & 0 & & 0 & \\
0 & -1 & 2v & -1 & & & & & \\
0 & 0 & -1 & 2v & & & & & \\
& & 0 & & \ddots & & & 0 & \\
& & & & & 2v & -1 & 0 & 0 \\
& & 0 & & 0 & -1 & 2v & -1 & 0 \\
& & & & & 0 & -1 & 2v & -1 \\
& & & & & 0 & 0 & -1 & 2v+\beta
\end{pmatrix},
$$

$$
\boldsymbol{\Delta}_p =
\begin{vmatrix}
2v & -1 & 0 & 0 & & & & & \\
-1 & 2v & -1 & 0 & & 0 & & 0 & \\
0 & -1 & 2v & -1 & & & & & \\
0 & 0 & -1 & 2v & & & & & \\
& & 0 & & \ddots & & & 0 & \\
& & & & & 2v & -1 & 0 & 0 \\
& & 0 & & 0 & -1 & 2v & -1 & 0 \\
& & & & & 0 & -1 & 2v & -1 \\
& & & & & 0 & 0 & -1 & 2v
\end{vmatrix}.
$$

© Springer Nature Switzerland AG 2020

N. Verichev et al., *Chaos, Synchronization and Structures in Dynamics of Systems with Cylindrical Phase Space*, Understanding Complex Systems,
https://doi.org/10.1007/978-3-030-36103-7

It can be established that, for main minors of the determinant, the following recurrent equation is valid [1]:

$$\Delta_k = 2v\,\Delta_{k-1} - \Delta_{k-2}, \quad k = \overline{1,\,p},$$

with boundary conditions $\Delta_{-1} = 0$, $\Delta_0 = 1$. In the cases that are of interest, $|v| < 1$. Under this condition, this equation has the following solution:

$$\Delta_k = \frac{\sin(k+1)\theta}{\sin\theta}, \quad \theta = \arccos v.$$

Consider polynomial $P(v) = \det \mathbf{A}_n$. Expanding the determinant of the matrix on the first, and then on the last line, we obtain the expression of the desired polynomial:

$$P(v) = (2v + \alpha)(2v + \beta)\Delta_{p-2} - (4v + \alpha + \beta)\Delta_{p-3} + \Delta_{p-4}.$$

Taking into account the recurrence equation, the polynomial is simplified:

$$P(v) = \Delta_p + (\alpha + \beta)\Delta_{p-1} + \alpha\beta\,\Delta_{p-2}.$$

If we assume in this equation that $v = 1 - \lambda/2$, then $P(v) \equiv P(\lambda)$ is the characteristic polynomial of the matrix, and $P(\lambda) = 0$ is its characteristic equation. Let us consider several cases of parameter values.

(1) $\alpha = -1$, $\beta = 1$. In this case,

$$P(v) = \Delta_p - \Delta_{p-2} = \frac{\sin(p+1)\theta - \sin(p-1)\theta}{\sin\theta} = 2\cos p\theta.$$

The characteristic equation has the form

$$\cos p\theta = 0.$$

By solving it, we obtain the eigenvalues of the matrix:

$$\lambda_j = 4\sin^2\frac{\pi + 2\pi j}{4p}, \quad j = \overline{0,\,p-1}.$$

(2) $\alpha = -1$, $\beta = 0$. In this case, we obtain the following characteristic equation:

$$P(v) = \Delta_p - \Delta_{p-1} = \frac{\cos(p+1/2)\theta}{\cos\theta/2} = 0.$$

By solving it, we obtain the eigenvalues of the matrix:

$$\lambda_j = \frac{4\sin^2(\pi + 2\pi j)}{2(2p+1)}, \quad j = \overline{0, p-1}.$$

(3) $\alpha = 0$, $\beta = 0$. In this case, $P(v) = \Delta_p = \frac{\sin(p+1)\theta}{\sin\theta}$. By solving it, we obtain the eigenvalues of the matrix:

$$\lambda_j = \frac{4\sin^2 \pi j}{2(p+1)}, \quad j = \overline{1, p}.$$

Reference

1. Belykh, V.N., Verichev, N.N.: Spatially homogeneous autowave processes in systems with transport and diffusion. Izv. universities. Ser. Radiophys. **39**(5), 588–596 (1996) (in Russian)

The manufacturer's authorised representative in the EU is Springer
Nature Customer Service Centre GmbH, Europaplatz 3, 69115 Heidelberg,
Germany. If you have any concerns regarding our products, please
contact ProductSafety@springernature.com

Printed and bound by CPI Group (UK) Ltd, Croydon, CR0 4YY
28/11/2025
02007692-0007